ICME-13 Topical Surveys

Series editor

Gabriele Kaiser, Faculty of Education, University of Hamburg, Hamburg, Germany

More information about this series at http://www.springer.com/series/14352

Ewa Swoboda · Paola Vighi

Early Geometrical Thinking in the Environment of Patterns, Mosaics and Isometries

Ewa Swoboda
Department of Mathematics and Natural
  Science
University of Rzeszów
Rzeszów
Poland

Paola Vighi
Mathematics and Computer Science
  Department
University of Parma
Parma
Italy

ISSN 2366-5947   ISSN 2366-5955   (electronic)
ICME-13 Topical Surveys
ISBN 978-3-319-44271-6   ISBN 978-3-319-44272-3   (eBook)
DOI 10.1007/978-3-319-44272-3

Library of Congress Control Number: 2016947510

Printed on acid-free paper

This Springer imprint is published by Springer Nature
The registered company is Springer International Publishing AG Switzerland

Main Topics You Can Find in This "ICME-13 Topical Survey"

The main topics you can find in this ICME-13 Topical Survey are:

- Theoretical background of research in early geometrical thinking;
- The role of early geometry in mathematical learning and understanding;
- From the shapes to the figures;
- From a static to a dynamic geometry: the isometries; and
- Rhythms, regularities, and patterns in the work with young pupils.

Contents

Early Geometrical Thinking in the Environment of Patterns, Mosaics and Isometries

1 Introduction

This survey confirms that one of the research trends in early geometrical reasoning has been a focus on creating a theoretical basis for research in this area. The main reason for the work on the theoretical foundations has been the fact that building geometrical concepts proceeds according to other regularities more than it takes place while building arithmetic knowledge. In this way, a broad research area was outlined. The first theory was created by Dina and Pierre Van Hiele (the Netherlands), but many on-going studies are related to this issue and many have focused on analysing and further expanding this theory.

In parallel, new conceptual approaches that capture the question of the theoretical basis in a different way have been created. They should be considered complementarily, as they repeatedly point to the other aspects of the geometrical knowledge. The examples discussed in this paper come from the circles of scholars gathered around Alain Kuzniak (France) and Milan Hejný (Czech Republic). A networking between Van Hiele's theory and the description of geometrical paradigms created by Kuzniak examine ways that children work at the lowest levels. This allows analysis of how they gain experience, which is the basis for the transition in the area of further paradigms. This is therefore a departure from the theories that have a linear structure and focus only on a successive development.

Milan Hejný indicates new aspects for descriptions of understanding of geometrical concepts, but first and foremost, he connects his proposals with the concept of schema-oriented education, where creating skills and procedures should be seen from a long perspective. In his theoretical background, a relatively new trend concerning the development of geometric reasoning is a pro-ceptual approach to the concepts. The formation of geometrical concepts is related to the empirical

© The Author(s) 2016
E. Swoboda and P. Vighi, *Early Geometrical Thinking in the Environment of Patterns, Mosaics and Isometries*, ICME-13 Topical Surveys,
DOI 10.1007/978-3-319-44272-3_1

abstraction. Assuming that the first geometrical knowledge is passive, comprehensive, global, and, therefore, static, conferring dynamism to geometric reasoning starts to be an important need.

This new trend should be interpreted broadly. For a long time, the role of manipulation in early geometrical tasks was not associated with reasoning but rather referred to children's working style, age, and ways of gathering information about the world. The specifics of children's work indicated the use of different items for manipulation. It seemed obvious that the children's work in active environments would more suitable for them, and functioning at the symbolic level would not be available. This style of work rather was implied by the Piagetian approach, which deals with the necessity of interiorisation of actions in the process of building mathematical concepts. However, it was not connected with the approach suggested by Gray and Tall (1994), which pointed to the necessity of joining processual with conceptual understanding. The pro-ceptual approach was hardly accepted by either the theory or practice of teaching geometry. Currently, designing a manipulative educational environment is focused on building a scheme for deep understanding of geometric concepts. The emphasis here is set not so much on observing objects in motion nor on the final results of manipulation, but on the ability to predict the result of the transformation.

In this survey, the research on the understanding of geometrical concepts has been grouped around two main issues: the understanding of geometric figures and the functioning of these figures in space.

The problem of understanding these figures that was indicated by Van Hiele's theory is still worth considering and research has brought much information on how it takes place in children's minds. The research reveals many limitations that give children trouble with the transition to higher levels of understanding. This research direction is dominant. Students' understanding of figures has been repeatedly carried out by analysing the classification of the figures they have made and their ability to describe objects and exclude counterexamples. On the other hand, research shows that children who are functioning in a static situation based on the recognition of a geometric object are doing it much less successfully than those that have the possibility of analysing the object given to them for manipulation.

At this stage of research, it is worth dealing with the level that has often been referred to as the zero level (earlier than the level that was described as the first level in Van Hiele's theory). Four to six-year-old children can and do successfully discover the world of geometry in many areas, but they do it in their own specific way. To use those early experiences for creating further stages of understanding, knowing, describing, and understanding them is required. Such research is performed too rarely. Additionally, this type of research requires specific methodology and being skilful in making proper observation and in analysing children's behaviour.

Another area of research has been the understanding of three-dimensional figures by children. It has been associated with criticism of the fact that dealing with solids takes place on higher levels of learning geometry. This goes against the natural way of discovering the world by children, which is made up of three-dimensional objects.

Much of the space in this survey has been dedicated to research related to regularities in geometry. This seems reasonable, as so far patterning in geometry has been treated superficially, mainly to determine whether a child can note suggested regularity. Researchers are of the opinion that the functioning of children in a world of regularity is important for not only their general mathematical development but also their geometrical.

In the research on children's understanding of regularity, one can distinguish a number of issues—from the very fact of their perception of regularity by creating arrangements of surfaces to geometrical relations hidden in mosaics. Following Steen (1990), mathematics is actually seen as a "search for patterns": "It is natural to try to find the most effective ways to visualize these patterns and to learn to use visualization creatively as a tool for understanding" (p. 3).

2 Theoretical Considerations About Early Geometrical Thinking

2.1 Why Early Geometry Is Related to Isometries

Geometry is one of the best areas for a child to enter the world of mathematics. The geometrical world can be opened very early because geometrical knowledge correlates very well with children's natural cognition. All the information gathered by perception has special importance for them. Learning for young children mainly consists of acquiring information by observing the world made up of objects. One of the features of these objects is their shape. Among shapes, there are regular ones. Perceived objects provoke further action. Children often say that a triangle is "very nice," which is another way of saying that it is regular (Hejný 1993). Since any regularity is attractive, it is easy to interest a child with it, and the motivation for any action is then natural. Various activities facilitate the learning of objects, creating intuition on which geometrical concepts are built. This is the main reason why geometrical concepts recognised by perception are closer to children's abilities than arithmetical ones.

One of the ways to further explore the world of geometry is to provide proper visual information associated with the possibility of manipulation and experimentation, with room for a child's own creativity and ingenuity. In a patterns environment, recognizing a geometrical concept is spontaneous and is connected with solving problems in which children are able to clearly define the purpose of their own actions. Such an approach is consistent with the psychological opinion concerning learning:

> A great part of our knowledge, as can be traced to our behaviour, is only implicit. We take up information with the help of invariants, without expressing or even able to express these invariants.... The cognitive analysis of such behaviours very often reveals the existence of powerful implicit mathematical concepts and theorems. (Vergnaud 1990, p. 20)

Accumulated experience enables children to create a *data set* that is used by them to build up their mathematical knowledge.

The use of materials for manipulation (e.g., stacking ready-made elements) has been considered to be the most effective learning environment. In addition, children's drawings also have a high research value (Swoboda 2007). Patterns—arranged with blocks, folded with puzzles, made from plasticine, lined from small pictures and figures, or drawn—are a friendly environment for children as they are close to their natural, spontaneous activity. They give the pleasure of creation without worrying about the outcome, create a chance to speak out without the fear of criticism, enable the realization of one's own ideas, and give motivation for manual and intellectual work.

Research conducted in the United States has shown that more than 94 % of children beginning their education are able to count to 10 and identify basic shapes (Ginsburg 2004). Additionally, children between 5 and 10 can act in the "world of regularities" by discovering them. During the creation of geometrical compositions, constructing buildings with blocks, or decorating carpets, children not only better learn geometrical shapes (by comparing the lengths of the sides or recognizing the size of the angles). They may also feel the need for such arrangements that an adult can describe using the language of geometrical relationships. Symmetry, illustrated either in a broader or narrower sense, is an idea that has been used by people to describe beauty, order, and perfection. These arrangements may appear accidentally, by trying and checking, until the child considers them to be sufficiently pretty. A sense of order tends to be verified visually by children. Hence, propaedeutic of geometrical figure-to-figure relationships may reside in the sense of a certain order or harmony—a specific arrangement of a surface or available fragment of space.

The idea of engaging children in the world of rhythms and regularities is a welcome phenomenon. The preschool period is already a good time for children to become interested in building shapes and finding patterns (Clements 2001). Generally, it has been stated that creating their own patterns is a good starting point for children's understanding of geometrical transformations. It is a long way to go from building a mosaic to creating geometrical concepts, but the connection between both is clear. In some handbooks for teachers, there are suggestions to do exercises with changing a figure's position, such as drawing patterns and mosaics where translation, rotation, and mirror symmetry are used (Jones and Mooney 2003). The creative process included in children's activities is regulated by perception. Theories concerning the development of geometrical reasoning stress that, at first, understanding is passive—consisting of attracting attention to a particular phenomenon: the shape of a figure or line or the mutual arrangement of objects in relation to each other. This cognition is static, stimulated by aesthetic feelings. However, geometrical rhythms have a special status. Continuing repetitive patterns is an activity that requires recognition and understanding of a structure and the ability to reproduce it. An inner structure consisting of the repetition of a group of elements suggests continuity or motion (Marchini 2004). Such motion can be described using the language of any geometrical transformation and can be recognized by children as a stimulus to any action (Marchini and Vighi 2011). Konior (2003), while writing about patterns, affirms: "Grasping of the rhythms exposed both in action and effects enables the student to embrace the whole sequence of extending and widening in one act. Therefore, it helps to gradually become free from the embarrassing limited model." (p. 36).

2.2 Impact of Visualisation on Geometrical Thinking

The epistemological problem of knowledge's origins in mathematics is strictly connected with the geometrical field. Geometry appears as a way of seeing the real world through mathematical eyes. However, the problem of being sensitive to geometrical phenomena is complex. Let's start from the quotation:

> The first, and the most basic, understanding of the real world is understanding via senses. We look at the world of geometry, but not with our eyes; we learn the world of geometry, but not with ordinary senses. Geometrical seeing is possible only because of the sixth sense. This seeing is not less obvious than seeing the real world using the sense of sight.... Those who lose geometrical seeing cannot approach the geometrical world; they can only listen to us, talking about this world. They are as the blind as those who find themselves in a gallery and listen to what the others say about the pictures. (Vopěnka 1989, p. 19)

At the beginning, there is neither geometrical world nor geometrical object in a child's mind. Only objects from the real world exist. However, we focus our attention on those objects in various ways. Vopěnka (1989) describes such a situation in the following way:

> To see "this" means to focus attention on "this" in order to distinguish "this" from the rest of the whole. This, what can absorb the whole attention on itself, we call a "phenomenon." (p. 19)

Perceiving "something" creates the *first understanding*. Children can focus their attention on the shape of an object or on the specific position of one object in relation to another. *Phenomena* open the geometrical world to a child. In spite of the fact that our attention is attracted by these phenomena, this first understanding is passive: stimulus goes from the phenomena. In this depiction, the role of perception is large: the perception of "something" is the first step to creating a child's own geometrical world (Vopěnka 1989; Hejný 1993). To make another step into the geometrical world, it is necessary to work in the physical environment: watching and touching the objects promotes the spatial, visual, and tactile experience; moving objects improves the concept of movement. At the beginning, it is important to promote a "handmade approach" to fundamental geometrical ideas, observing properties and, in particular, invariants. During such activities, children create *pre-concepts*.

If we accept the fact that this view is of significant importance to the first level of geometrical cognition, we also have to consider psychological provisions concerning cognition. The results of psychological research (Kaufman 1979) confirm that, in the process of grasping shapes, pictorial designates are of great importance. In addition to this, dominance of the whole over the part is a regularity in perceiving shapes. The rules of structuring an image investigated in view of the information analysis system suggest that regular, symmetrical forms and shapes are the most easily recognized, as one element can be predicted from another (Grabowska and Budohoska 1992). Regularities, groups creating some logical wholeness, can be elements of a composition regulated through visual perception. Jagodzińska (1991) writes:

> There are well-known arguments of gestalt psychologists who proved that the perception of shapes and objects are of a primal character while discerning constituent elements is the outcome of further analysis. As a matter of fact, we can quote here some interesting data that suggest that, indeed, in the perception process sequence, the global structure of an image precedes detailed analysis. (p. 64)

Demidow (1989) gives a broad survey of the research conducted by physiologists concerning the mechanisms of shape recognition. We can also find information there about invariant transformations conducted by our eyesight. For example, pictures of different sizes are invariant (unchangeable) to the organ of sight (the eye identifies them); this is the same when the position of an object is changed—but only up to 15°. The mirror image is not invariant, even though children are born with such properties of perception; as humans develop, the eye loses the invariance of mirror images.

These remarks have an essential meaning in geometrical environments; they are referred to as *patterns*. Creating bands or mosaics was unequivocally assessed by Van Hiele as operating on a visual level that did not require the internalization of actions. He refers to the structures of the first level as visual, *structures of the appearance*; they are manifested in recognizing regularities or certain wholeness. According to this theory, all perceived regularities are classified as visual structures. The things that inspire children, propel them to action, and undergo control and that they reflect upon are rhythm, order, and regularity. Such action seems to be in accordance with the original meaning of the Greek word *symmetros*, which meant "harmonious" or "well-proportioned." This interpretation resembles the assessment of the mosaics that have been created by humans since the earliest days of their history. It seems that the recognition of a specific figure-to-figure position is only a static image of this relation and is not connected with the movement of one object onto the other.

This leads us to the conclusion that in situations where balance, stemming from an appropriate arrangement of the elements that constitute an image, is present, there is no need to introduce movement. Children working in an environment of visual regularities do not appeal to the idea of movement, placing one object onto the other. The understanding of relations between figures as dynamic arrangements of space is placed, so to say, on the opposite pole. Acts of perception are important, but they are not a sufficient source of geometrical cognition. Szemińska (1991, p. 131) states that perception gives us only static images; through these we can only catch some states, whereas by actions we can understand what causes them. It also guides us to the possibilities of creating dynamic images. During manipulation, children's attention should be focused on the *action*, not on the *result of action*. This requires a different type of reflection than the one that accompanied their perception. The process of acquiring such skills is lengthy and gradual (Szemińska 1991). The work on geometrical transformations involves both static and dynamic aspects: it supports the reasoning and the flexibility.

2.3 Theoretical Background of Research in Early Geometrical Thinking

It is a common opinion among researchers that the formation of geometrical concepts takes place in the different way than it does in the formation of arithmetical concepts.

Tall (1995) has formulated a theory on how an individual builds up mathematical ideas. He has distinguished three main sources for creating mathematical concepts: perception, action, and reflection. These three sources are the basis for three essentially different kinds of mathematics:

- *space and shape* (geometry), based on theorizing about the (geometric) objects: we perceive and construct at increasing levels of sophistication;
- *symbolic mathematics* where actions on objects (such as counting) are symbolized giving new mathematical concepts (number); and
- *axiomatic mathematics*, built by reflection on the properties of the first two forms of mathematics in terms of formal definitions and logical deductions (Fig. 1).

Fig. 1 Various types of mathematics, from Tall (2001)

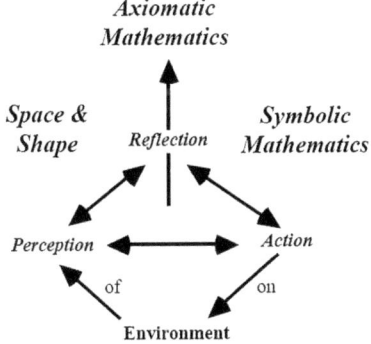

Perception is therefore the primary source for creating the geometrical concepts. Regardless of this, there are theories that describe the process of formation of the geometric concepts and geometric reasoning. They are the base for diagnosing the degree of formation of mathematical knowledge. References to these theories are also found in the didactical proposals.

A short survey of theories on creation of geometrical concepts should start from Dina and Pierre Van Hiele's findings. *Theory*, published in 1957 in Utrecht, described in detail in the publication *Structure and Insight*, is one of the most well-known and frequently used bases for analysis in teaching and learning geometry.

The main feature of their theory is that they have determined the levels of understanding of geometrical concepts. These levels define the hierarchical structure of building knowledge. According to the authors, it is not possible to revolve any of the levels—achieving higher levels is possible only after mastery of the previous level. The levels, as defined by Van Hieles, are as follows:

- Visual level
- Descriptive level
- Relational level
- Deductive level
- Rigor.

During the first level—the visual level—concepts develop on the basis of experiences and conscious observations from reality: students first learn to recognize shapes then analyse the properties of the shapes. The visual level is the main step in spatial knowledge. On the visual level, students recognize a figure as a whole and are able to represent it as a mind vision. Note that Van Hiele (1986) states that "the levels are situated not in the subject matter, but in the thinking of man" and Arcavi (2003) suggests that visualization can be considered as a method of "seeing the unseen." Moreover, Viholainen (2006) states that "*visual thinking* is probably the most usual type of informal thinking in mathematics."

At the visual level, therefore, the student:

- identifies, compares, and sorts shapes on the basis of their appearance as a whole;
- solves problems using general properties and techniques (e.g., overlaying, measuring);
- uses informal language;
- does not analyse in terms of attributes.

Later students see relationships between shapes and make simple deductions. Only after these levels have been attained can they create deductive proofs.

Van Hiele did not deem that any of the levels was free from the thinking. In particular, it cannot be assumed that the visual level eliminates action (manipulation) by objects. De Lange (1987), presenting his interpretation of Van Hiele's theory states that "a pupil reaches the first level of thinking as soon as he can manipulate the known characteristics of pattern that is familiar to him" (p. 74).

Research conducted mainly in Valencia has elucidated this theory. They are related to many different aspects of geometric activity: recognizing figures, drawing, use of terminology and verbal description, the logical identification of relations, and the ability to apply concepts. First of all, the researchers noted that the levels of Van Hiele's theory were not discrete, so a more in-depth study of the transition from one level to another was needed. It was found that, among other things, it was necessary to more accurately define the "contents" of each of the levels. Students can function at a given level for a long time: does this mean that their knowledge at this time does not change? This has led researchers to distinguish a category called degree of acquisition that, in their opinion helps in didactical research (Fig. 2).

The degree of achievement of a given level is determined by observing how children work and on the basis of trying to determine their ways of thinking. So on the *no acquisition* step students are not aware of or do not feel the need for ways of thinking specific to that level. Student at the *intermediate acquisition* level will use these methods often and consistently, but in difficult and unusual situations they will tend to return to a lower level. The work carried out by this group of researchers (Gutiérrez and Jaime 1998; Gutierrez et al. 1991) provides an opportunity to determine the conditions of transition from one level to another (higher) one. It must also be taken into account that it is possible to simultaneously achieve two different levels of understanding.

Van Hiele's theory has influenced trends in research on the formation of students' geometrical knowledge, but has also strongly narrowed the examination of early geometry. The model has greatly influenced geometry curricula throughout the world by emphasis on analysing properties and classification of shapes at early grade levels (e.g., associated with classifying triangles or quadrilaterals).

Regardless of the fact that this trend is still valid, some researchers have attempted to go beyond the framework set by the model created by Van Hiele. Criticism of the theory not only refers to the narrow treatment of levels (a common complaint is the omission of the level of the early formation of geometrical concepts). Investigators have criticised the theory as being "too linear" and focused only on successive development.

Hejný and his team have carried out work that has extended and complemented this theory. Hejný has focused on the emergence of the geometry world from reality and on building the subsequent stages of understanding of concepts. These are also

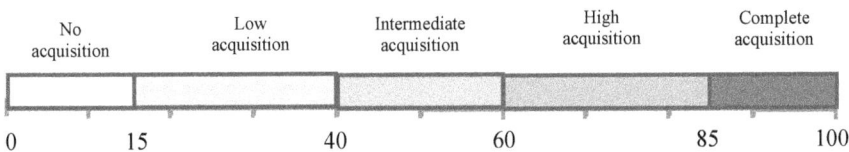

Fig. 2 Degrees of acquisition of a Van Hiele level (Gutierrez et al. 1991, p. 238)

the levels of understanding, but they show the evolution of concepts on larger spectrum. Hejný built his theory based on both experimental studies and the conclusions resulting from the theoretical descriptions from other authors such as Piaget, Vygotsky, and Van Hiele. In his theory of the development of understanding, the geometry of the world passes through levels. Very characteristic of this is the precognitive level, where shapes are understood as attributes of real objects. The author gives three criteria characterizing this level of understanding of geometrical concepts:

- Among all the attributes, the child recognizes a special class of attributes: shape, which is parallel to the classes of colour, taste, and quantity.
- Each shape as "a square," "a circle," "a rectangle," or "a cube," is associated with a collection of objects from the real world. Despite the ability to use names such as *triangle* and *pyramid,* descriptions such as *long* and *tall*, and even the ability to make certain types of comparisons such as *longer* and *wider*, they are still words and concepts related to the real world.
- The child does not admit the status of the shape of the object itself as existing independently.

At this level of understanding, children treat the drawing of a geometric figure as a shape that must be completed: a circle can be an unfinished drawing of the sun or a baby's mouth; a square can be an unfinished drawing of windows, a block, or the outline of a house (Hejný 1993).

The next level takes place when children start to perceive the same shape in a variety of subjects and when attention shifts to the shape as such. This is the independence of geometric phenomena that is strengthened by assigning a separate name, such as *square* or *circle*.

Recent research by Hejný and colleagues has been related to the implementation of his theory to practice what is called *scheme-oriented education*. The entire idea of scheme-oriented education (especially on an elementary level) is based on the assumption that most knowledge (including mathematical) is gained not through focussed learning but through repetition of various life experiences. They have highlighted two main issues: the long-term building of mathematical concepts and procedures and connections between mathematical concepts and experiences that a child gets in everyday situations. Another feature of this approach is connected with the distinction between process and concept as described by Gray and Tall (1994). Hejný (2012) showed the importance of perceptual transfer in a pupils' minds when they are grasping a processually perceived situation conceptually or a conceptually perceived situation processually. It is the latter of the two directions that is much more frequent in geometry than in arithmetic (Jirotková 2016). For this reason, many educational proposals created by Hejný and colleagues take place in the physical, manipulative learning environment. Children do not only play with the models of the figures and describe them, but also solve proposed tasks that require reasoning.

However, a more popular approach in the research community has been to depart from a linear description of the development of the understanding of geometrical concepts. One of these approaches has been concerned with the different paradigms of geometry and has been developed by Kuzniak and colleagues.

In describing his approach, Kuzniak states:

> The geometrical world representations of objects often remain spatial objects. and, in fact, the way is very long from a real spatial object to the notion of 'figural concept' described by Fischbein (1993). He drew attention to the fact that the development is carried out by scientific revolutions that replace the old paradigms with new ones. Our research puts in evidence three different paradigms that bring us to distinguish various forms of geometry. To clarify these paradigms we used the forms of knowledge of space put in the evidence by Gonseth (1945–1955): intuition, experiment, and deduction. We revisited them in the light of recent contributions to the historiography of mathematics and also in a perspective of teaching, which gives a different view of this knowledge (Kuzniak and Houdement 2001).

Recent activity in primary education has involved working within the paradigm of Geometry I. Here is how it has been described by Kuzniak:

> Geometry I (Natural Geometry). "The source of validation is the senses. It is intimately related to reality. Intuition is often assimilated to immediate perception, and experiment and deduction act on material objects by means of the perception and instruments. The backward and forward motion between the model and the reality is permanent and allows proof of assertions. For example, dynamic proofs are accepted in this Geometry".

In this approach, a reference to active solving of geometrical problems is apparent, not only to recognising objects. Active problem solving allows practical operations—including construction (of physical objects), drawing, and visual verification. An example of this is the exploration of triangles constructed using sticks of different lengths and then determining when the construction is possible and when it is not.

An alternative to this is Geometry II (Natural Axiomatic Geometry).

> The source of validation bases itself on the hypothetical deductive laws in an axiomatic system. A system of axioms is necessary but the axioms are as close as possible to the intuition of the space around us. The axiom system can be incomplete, but the demonstrations inside the system are necessary for progress and for reaching certainty. (Kuzniak and Houdement 2001, p. 4)

The last direction of his research has been aimed at describing the Geometric Working Space, focussing on the application of theory in practice. It is a multi-dimensional description of space in which geometric knowledge is built by students. Kuzniak has drawn attention to the need to combine different elements, such as a real and local space as material support with one set of concrete and tangible objects such as figures or drawings, a set of artefacts such as drawing instruments or software, and a theoretical reference system based on definitions and properties (Kuzniak and Nechache 2015, p. 544). Because knowledge is built by its users as a human activity, it is necessary to consider another dimension: the cognitive one, which includes a process of visualization related "to the representation of both space and material support, a process of construction and function of the instruments used (e.g., rulers, compasses) and the respective geometrical configurations, and a discursive process producing arguments and proofs" (Kuzniak and Nechache 2015, p. 545).

Recently, research in mathematics education has turned its attention to the problem of "language and semiotic aspects in the construction of mathematical knowledge, both in an individual and in a social construction perspective" (Boero and Consogno 2007). In particular, starting from the assumption that the language is fundamental since the mathematical objects are not directly accessible Duval (1993), elaborated a theory based on this main idea: the learning of mathematical objects is necessarily conceptual and an activity on them is possible only using 'registers of semiotic representations' (Duval 1993). He provides a very rich theory about it based on the assumption that there is no knowledge without representation. Moreover, two kinds of transformations are mathematically relevant: the "treatment" (Duval 1993, p. 41), the transition from a representation to another in the same register, and the "conversion" (Duval 1993, p. 421), the transition from a representation in a register to another in a different register. The transition from a semiotic representation to another and vice versa is essential for the conceptual learning of mathematical objects: "Thinking in mathematics depends on the synergy of several registers and not on the activity of a single system. Unlike what occurs in other fields, mathematical concepts are only understandable within such a synergy" (Duval 2006, p. 21).

Duval (2005) affirms that geometry requires a cognitive activity very complex but complete, since it stimulates the gesture, the language, and the seeing. It is a field of knowledge that implies the cognitive joining of two very different representation registers: the visualization of the shapes and the language; a synergy between visualization and language is fundamental to understand geometrical arguments.

Duval (2005) identifies the origin of the difficulties in geometry in the intuition which relies on perception. According to psychological studies, perception plays a fundamental role in the visualisation process: "By perception the visual thought organises itself as a starting point of insight and reflection, mental activities which contribute to the formation of concepts" (Marchini et al. 2009, p. 62).

Perception is a process of selection and organization, cognitive activities connected with knowledge and understanding. Nevertheless, visual perception may hinder the way of seeing geometrical figures. Following Duval (2005), this way depends on the activity in which it is involved. There are two ways of seeing a figure: iconic and not iconic. The second is a sequence of operations of geometrical property identification that implies that the "deconstruction" of the shapes has been visually recognized.

This brief overview of theories that underlie geometrical knowledge has been focused mainly on those elements that affect research conducted at the lowest educational level. However, even such a selective approach shows that theoretical background can be varied, taking into account many different aspects of building the geometrical knowledge of students. It also shows the specifics of the geometric research that distinguish it from research in other areas of education.

3   Early Geometrical Thinking in Research

In children's geometrical intuition, there are at least two fundamental areas of geometrical concepts developed during school education. These are:

- identifying and creating shapes (circle, square, triangle, etc.) and
- creating arrangements based on a variety of regularities, symmetries, and repetitions.

Both areas are intimately linked, because shapes occupy a specific location (position) in space.

3.1   Research About Identifying and Creating Shapes

3.1.1   Introduction

Traditionally, early education has been associated with children's detection of geometric figures and with the ability to reproduce specific shapes; this has been one of the dominant trends in research. Children from an early age meet with various shapes through contact with surrounding objects. When looking at them and getting to know them with their other senses (e.g., touch), they are able to focus on one of their properties: shape. Real objects—solids—have shapes and children focus their attention on this attribute. Such first recognition is passive and comprehensive, is constituted by acts of perception, and is does not involve logical justification or analysis of the properties of the shape.

Because everyday language contains terms that can be associated with the concepts and geometric properties, children are able to use the names of geometric figures, such as triangle, circle, and square, relatively early. This does not mean, however, that they use geometric concepts. Clearly, the use of such vocabulary is not a basis for determining the degree of mastery of geometric concepts. These words that are the names of geometric figures are related to a specific objects known to them: a plate, the wheel of a bicycle, a window, a road sign, a roof, etc. They use them just like other adjectives to describe the objects of the real world. These terms are therefore only attributes of real objects.

Hejný (1993) proposed the following characteristics for figures in the school, whose understanding is determined by two features:

- number of parameters and
- the position of the figure in relation to the vertical-horizontal direction.

In this approach, a circle has only one parameter (radius) and there is no reference to the vertical-horizontal position. A square has a one-parameter status, but it shows strong relationships with the system of vertical-horizontal direction: it is easiest to put the sides of the square in line with the vertical level, it is harder when

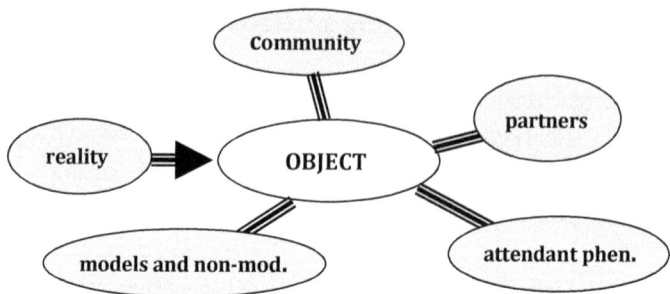

Fig. 3 Scheme of relationships among geometric objects (Jirotková 2016)

the diagonals are in line with the vertical level, and the most difficult is the general. An equilateral triangle has one parameter. A rectangle has two parameters, like the rectangular or isosceles triangle. Any triangle has three parameters, which shows the complexity of the concept of a triangle (Jirotková 2010).

Deepening of understanding of geometric shape is done by extending compounds in which it operates. In comparison with other figures (as well as when presented with examples and counterexamples), a child realizes properties of figures —attendants—(visual and hidden) and learns to describe these compounds. The following diagram presents the contexts that understanding a figure depends upon (Fig. 3).

3.2 Results of Research on the Understanding of Geometric Figures

3.2.1 Understanding of Two-Dimensional (2-D) Figures by Children

Research aims have usually been associated with qualification of the level of students' understanding of basic geometric shapes: triangles, circles, and squares. The problem that has appeared repeatedly in the research has been the problem of students' understanding of the triangle (Clements et al. 1999; Levenson et al. 2011). These studies have often relied on analysis of how students classified figures from the shapes presented on a board. One of these boards is present in Fig. 4.

Investigations may also be focused on the ability to separate squares and circles from the non-examples. Such studies have confirmed that even 3–6 year-old children are able to extract from a collection of many shapes those that belong to the same class. The criterion of their actions is the visual assessment of the shape, without going into the properties that define the shape. Another factor determining their choice was typicality of the shape (during research between typical rectangles and triangles, very narrow rectangles or scalene triangles were placed). Research into the understanding of the concept of triangles (Hejný 1993; Tsamir et al. 2015;

Fig. 4 Students mark
triangles—a research tool
(Clements and Battista 1992;
Clements and Sarama 2016)

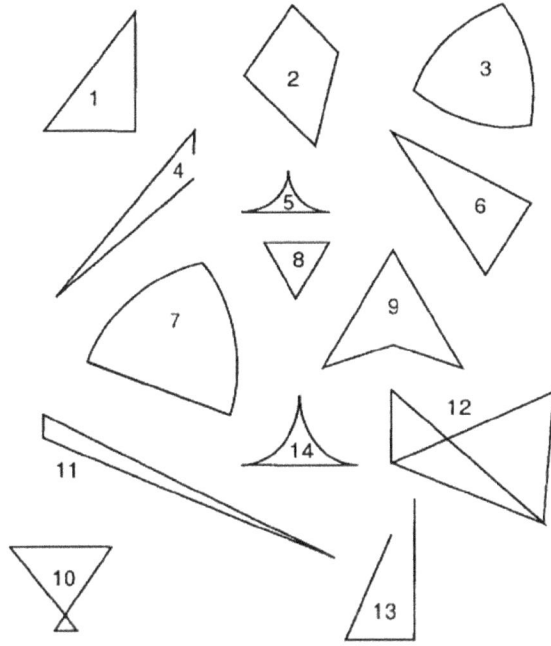

Vighi 2003a, b) shows that, in this case, the primal understanding refers only to an equilateral triangle.

Studies in which children were deprived of the possibility to manipulate objects has determined that their understanding of the figures is mostly at the visual level. Results of other investigations (Satlow and NewCombe 1998) indicated that children ages 3–5 rejected more of the typical figures than the invalid figures (open triangle-like shape). When reviewing the children's descriptions of circles, triangles, and rectangles, only a few children referred to the attributes of these shapes, indicating that most children were operating at the first Van Hiele level of geometrical thinking (Levenson et al. 2011). These students underwent further study, and as a result they were also shown to be capable of recognizing components and simple properties of familiar shapes. The authors of these studies conclude: "Thus, evidence supports previous claims (Clements and Battista 1992) that a prerecognitive level exists before Van Hiele Level 1 ("visual level") and that Level 1 should be reconceptualised as syncretic (i.e., a synthesis of verbal declarative knowledge and imagistic, each interacting with the other) instead of visual" (Clements et al. 1999, p. 192).

Observations of children who actively investigate shapes (they have the opportunity to manipulate them while solving some problems, including classifying shapes) provide other results. Children are able to build "families" for geometric and out-of-geometric shapes, taking into account such properties such as the number of sides, the convexity, and the complexity shape with curve and lines.

Various activities related to the manipulation of shapes affect their better recognition, and the need to determine the properties common for a "family" affects a better familiarity with shapes. This situation forces the use of language, which supports the transition to the descriptive level. In this way studies show that for children from the age of 5–7 years old in an active environment, evolution of the visualization is accessible for the students, although it does not seem to be used spontaneously (Coutat and Vendeira 2016).

Situations that involve naming of particular shapes and the influence of names to deal with figures have appeared in many publications. There have been observations that show that the fact that the word *triangle* can refer to many objects of everyday life can be an obstacle to the construction of the concept of triangle (Vighi 2003a). Coutat and Vendeira (2016) have noted that children tend to give shapes— both geometric and non-geometric—names that are associated with objects from the real world. Naming objects in this way makes it easier to deal with shapes as such.

Another problem that has often been undertaken relating to the understanding of geometric figures at a slightly higher level has been the study of the understanding of the relationship between triangles and quadrangles (i.e., classification of quadrangles). This problem can be placed at the transition between the descriptive and relational level. Research has focused on this problem due to the belief that students at this educational stage need not only know geometric properties but also need to understand relationships between properties and shapes. The aim of such studies has been to determine the ability to describe the figures and to observe attempts to create definitions. Studies have repeatedly been carried out that use models and figures that students have to group according to properties they choose themselves. Quadrilaterals and the relationships among them have often been a part of elementary school mathematics curricula. Research has suggested that students initially focus on visual characteristics of figures instead of their properties (Mack 2007). One of the reasons for having students operate with prototypical figures is their static understanding in the typical position: considering quadrilaterals to be static figures with certain properties (e.g., a trapezium is a figure with one pair of parallel sides, one of which is parallel to the bottom edge of the paper). This has been considered to be the main reason for students not being able to conceptualize the interrelationships among different quadrilaterals (Walcott et al. 2009).

Some of this research has analysed these skills while examining the impact of teaching style. A constructivist approach to teaching has developed the concept of inquiry-learning through self-exploration. Researchers have been trying to determine to what extent questioning may influence an increase in competence in describing figures (Lee 2016). The conclusion that has come from these studies suggests that when students are faced with the need to answer teachers' "why" questions they are forced into deeper analysis of the properties of figures. However, this generalization may be wrong. It is necessary to take into account the different needs of students, including different ethnographically embedded learning styles and possible different interpretations of questions. This has been a particularly significant problem in certain Asian countries (Hsu and Lin 2009).

3.2.2 Understanding of Three-Dimensional (3-D) Figures by Children

Although small children have a lot of experience with handling blocks through their typical play, early geometry has often skirted this range, moving the issues related to 3-D figures to higher educational levels. This situation has been recognized as worrisome. There have been voices that have raised the need to extend geometry to kindergarten with well-designed activities with 3-D figures in order to provide students with an initial understanding of spatial concepts (Sinclair and Bruce 2015). Regardless of the fact that teaching programs in many countries differ significantly (e.g., in Taiwan, tasks involving counting blocks in a 3-D figure are usually included in the first grade textbook), there is still much to explore in how to support children's knowledge about the geometrical aspects of solids. Still, too little attention has been paid to how to introduce children to understanding and solving problems involving 3-D objects. Conceptual knowledge of geometrical solids, visualizing these objects, and concrete activities with tangible materials or drawings seem to be closely interrelated. One of the research problems may be concentrating on ability to visualize multidimensional objects and presenting them on the plane. "As they construct block buildings during play, children deal with geometrical congruence, they distinguish solids according to the properties, or they recognize solids. Usually, this happens unconsciously and without any explicit naming of the solids but in connection to mental reflections on spatial relations, orientations or the structure of three-dimensional arrays.... [Because of] children's limited drawing skills at the primary age, ... we derive only limited information about children's geometrical knowledge of solids" (Reinhold and Wöller 2016, p. 2).

Some studies, such as Battista and Clements (1996), have suggested that children's skills in this regard are quite low. Other studies have analysed skill in describing the 3-D figures by students pursuing existing curricula. The problems examine have been determining the locations of 3-D objects relative to each other, recognizing 3-D objects, recognizing the properties of 3-D objects, constructing the 2-D and 3-D relationship, calculating the volume and area of 3-D objects, and recognizing the properties of 3-D structures comprised of identical objects (Denizli et al. 2016). They have shown that these skills increase with age, but it has been difficult to determine to what extent this is the result of either teaching or the natural development and experience gained from everyday activities.

Comparative studies on the ways students of different cultures describe a construction created from wooden blocks have shown similar acquaintance with their properties despite difficulties in describing the exact properties of the objects created (Reinhold and Wöller 2016). This also shows that the development of such knowledge is not culturally dependent and is associated with natural development of skills.

In this context, it is worth referring to studies involving specially designed lines of teaching that have focused on developing children's understanding of 3-D figures (Kloboučková et al. 2013). These findings illustrate how a pupil can build problem-solving strategies and how knowledge is spread among pupils.

In an environment of blocks, a didactical path has been created that enables the transition of spontaneous experiences into mathematical knowledge. This has been associated with the formation of language and the use of visual representation. The idea here has been the use of different forms of representation of this type of building, which could be:

(1) a physical model consisting of a real building constructed from blocks cube models;
(2) a portrait that is either hand-drawn, computer-drawn, or a photo of the physical model;
(3) a dotted plan where dots are inscribed into the squares of the building's floor projections; or
(4) a triple projection where a cube building is represented in three orthogonal projections: from above, from the front, and from a side.

These four representations capture the finished building, i.e., the concept. The last proposal is related to the process of building the construction using special symbols: $\square$ = put down cube, $\leftarrow$ = go west, $\rightarrow$ = go east, $\downarrow$ = go south, $\uparrow$ = go north, $\equiv$ = go up one floor (Kloboučková et al. 2013, p. 986).

Proposals for activities using computers have increasingly appeared. Using computer programs, children can manipulate blocks in virtual reality, building different constructions. Many researchers have maintained that computers have become an essential part of mathematical education (Van Heuvel-Panhuizen and Buys 2004), where the combination of activities in the real, physical world and the virtual world gives child the ability to collect experiences that help connect 2-D and 3-D representations. On the screen, children can make manipulations: rotations, translations, deletion of blocks from an existing building, approximations, and enlargements. After that, they can see the effects of their actions. Therefore, researchers have focused efforts on developing virtual manipulatives, creating a new trend of designing technology-integrated mathematical instructional materials. With this development, geometric objects can be created in an interactive environment to support multiple representations. Such expectations have been confirmed in the research outcomes (Yuan 2016). Students faced with the problem of determining the number of hidden blocks in a construction presented visually as a 2-D picture had great difficulty in their determination. Manipulation using physical wooden block was insufficient as the support: students were not able to merge two different representations. This problem has occurred less frequently among students working in virtual reality.

3.2.3 Teachers' Understanding of Basic Geometric Concepts

Starting from the obvious assumption that teachers' mathematical knowledge has an impact on students' knowledge, there have been several studies attempting to determine the level of teachers' knowledge. In general, since participants in these studies have been pre-service teachers, it has unfortunately turned out that most students entering pre-service teacher training programs have a very poor background in geometry, with many gaps and partial or erroneous knowledge. There have been examples in the literature showing that understanding of basic geometric concepts among teachers is not sufficient (Cilavdaroğlu 2012; Yenilmez and Yaşa 2008). Pre-service teachers could not comprehend the relationships between basic geometrical concepts. Some of studies have relied on examination of knowledge of basic definitions, such as the definitions of angle, polygon, triangle, rectangle, trapezoid, parallelogram, rhombus, oblong, square, deltoid, and circle. Very often the concept of trapezoid was examine (Fujita and Jones 2007; Usiskin and Griffin 2008; Türnüklü et al. 2013; Ozdemir and Dur 2014; Brunheira and da Ponte 2015). This can be observed in regard to the formal definition of trapezoids and general mathematical assumptions (Fig. 5).

Questions Q1 and Q2 shown in Fig. 4 are related to learner's knowledge of terms *image* and *definition*.

The education process these pre-service teachers have previously received has often been blamed for this situation (Usiskin and Griffin 2008), especially in regard to the very narrow presentation of this concept in curricula and school manuals. It should be stressed that a typical (prototype) example has been the key factor for the further use of the concept (Türnüklü et al. 2013). It has been common for pre-service teachers to use their own image of figures to create their definition (Fujita and Jones 2007). Additionally, research results show that they have actually defined prototypical images (Ozdemir and Dur 2014).

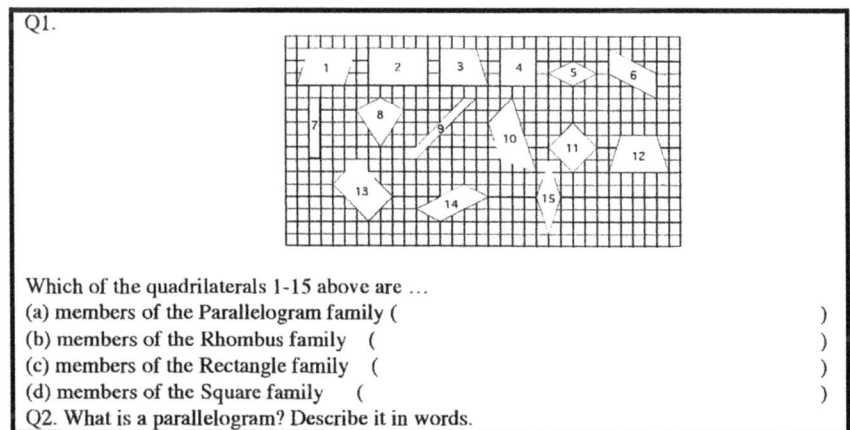

Fig. 5 A question from the research questionnaire used by Fujita (2010)

3.3 Didactical Proposals for Teaching the Concept of Geometric Figures

There have been relatively few descriptions in the literature of training proposals directed to children rather than teachers.

It is worth mentioning here the concept of geometric education of children that was created in the Czech Republic (Hejný 2012; Jirotková 2010; Hejný et al. 2007). It presents a global approach to the problem of geometric education that is focused on the understanding of plane figures, spatial intuition, and transformations. Based on analysis of Gray and Tall (1994), the idea has been subordinated to the concept of a *procept*, in which two cognitive principles—process and concept—are connected. In geometry, the most important and difficult task is the transition from a static to a dynamic approach, which is a conceptual situation perceived processually. A series of tasks, from very simple ones based on a global basis shape to complicated one requiring the use of properties of objects, have been designed in a very simple, child-friendly learning environment. However, it has been an environment in which the children are confronted with a problem requiring them to make a series of manipulations: a manipulative learning environment. In addition, it is a rich environment that refers to Wittmann's "substantial learning environment."

Here is an example of two extremes of tasks in a "stick environment":

Task 1. Move one stick so as to obtain a square.
Task 2. Eleven sticks built 5 triangles. Build 6 triangles with 12 sticks (Fig. 6).

Task 1 refers to a visual representation of the square—this figure can be easily recognized based on the mental representation of shape. Psychological foundations of perceiving shapes confirm that the eye tends to close lines. This task appeals to manipulation, but the solution is carried out at the visual level. To solve it, it is enough just push one stick to actually obtain the requested object.

The second task refers to a different mechanism. Here, the child after the first analysis of the proposed system can only note that in this construction the sides of some of the figures are common. Therefore, it was enough to have 11 sticks when

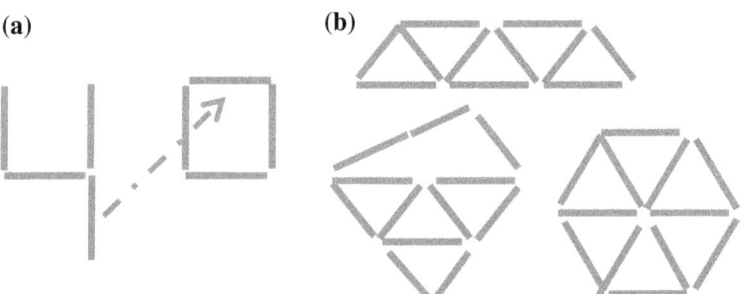

Fig. 6 a Task 1 and its solution. **b** Task 2 and its possible solutions

creating five triangles. Simply adding one stick to the existing configuration is not a good approach: it can create an "open triangle" or it can be used to divide one of existing triangles into two halves. The last action is in contrary to the instructions and leads to two new triangles instead of one. It is therefore necessary either to redesign the existing system in order to create a situation where more sides will be shared or to move away from the focus on equilateral triangles. The task therefore requires redesigning the prior scheme of thought associated with the task. When solved in the classroom, it provides an opportunity to experiment, verify ideas, discuss, argue, and develop a geometric language.

Equally interesting are proposals to build schemes concerning the functioning of objects in space, their mutual arrangement, visualizations, and relationships. An example is the following task: Create two buildings of four cubes such that by relocating one cube in one building you create a second building.

This task is open and allows creativity at many levels. The first task involves building an object with four blocks. Work on the second task can be in accordance with a child's own strategy. If the results are discussed in class, checking the solutions will require a transition from a static representation of both objects to imagining (identifying) the movement that caused the change of one object to another. This is therefore not the same as the observation of movement, but relies on trying to find the dynamic relationship between the initial state and the final one. It does not matter that this relationship cannot then be realized by doing this movement.

Another proposal associated with an investigation of active shapes involves combining science with art. Abstract paintings, with their features such as shapes, colours, and composition, can provoke a way of seeing the world through mathematical eyes (Vighi 2015). Creating or "reading" the art composition in a natural way places figures in space, causing figures and their mutual arrangement to be viewed simultaneously.

In recent times, the diffusion of communication by the mass media has led researchers to study the role of different languages and systems of representation as new important aspects in education. Art education has been present in Italian primary school curricula since 1985. The main idea has been that artistic culture has a formative role, as has been documented by research on art and perception. In particular, the *Italian National Guidelines for the Curriculum of Kindergarten Schools* (MPI 2012) states:

> The encounter of children with art is an occasion to observe the world that is around them with different eyes. The materials explored by the senses, ... the observations of works (paintings, museums, ...) help to improve the perceptive capabilities, to cultivate the pleasure of the fruition ... and to approach the culture and artistic heritage. (p. 20)

In other words, while previous attention had been focused on "production" (spontaneous drawing, etc.), now "fruition" was also important: looking not is a passive action but a dynamic activity of selection of shapes, colours, and configurations.

An important concept involved in a painting, starting from its planning to its realization, is the "concept of space." The canvas is an empty space that must be organized by placing objects (independent space). In other geometrical situations, the figures create the space (non-independent space): "... essentially, or primarily, we think that to the objects (or to the shapes), the space is only a coexistence of them" (Speranza 1997, p. 130). Usually, the space managed in a painting is a "microspace," namely a space that is manageable with the hands and the eyes (for instance, a sheet of paper). Sometimes it is also a "mesospace," manageable only with the eyes (i.e., a wall in a room).

In a space, the geometrical relations that describe the position of one object relative to another are more important than the figures as such. In order to demonstrate this, mosaics and tessellations, including those created by Escher, have been used. Although Escher's mosaics have often been used at higher educational levels, it turns out that children at lower levels can also successfully deal with them (Marchini and Vighi 2009a, b). Working with mosaics indicates that children pose the intuitions of geometrical isometries, but their development requires a conscious educational treatment. In particular, among the isometries, axial symmetry is a very complex topic. Research has documented the difficulties observed in its understanding. Piaget and Inhelder (1947) pointed out the individuation of a "vertical axis" of symmetry in very young pupils. This could be a didactical obstacle, as Brousseau (1983) has shown. Swoboda (2011) highlighted the difference between a static or a dynamic approach to axial symmetry among 4–6 year-old pupils. She designed an experiment on the construction and deconstruction of a pattern using printed tiles. Firstly, the tiles were "equal," having the same orientation. She then placed a "symmetrical tile" in the pattern. She observed that when children were requested to reconstruct the regularity of the pattern, they tried to rotate the new tile instead of turning it over. This indicates that the visual representation of the static relationship between objects is insufficient for a full understanding of isometrics as transformation.

3.4   Empirical Research about Creating Arrangements Based on a Variety of Regularities, Symmetries, and Repetitions

3.4.1   The Role of Patterns in Mathematical Development

The opinion that mathematics is based mainly on identifying and analysing regularities has functioned almost from its beginnings. According to some historians of this discipline, mathematical relations could already be identified in the geometrical decorations of items created by late ice-age humans (Encyklopedia Szkolna, Matematyka 1990, p. 140). Kordos (2005) states that:

It is worth paying attention to the richness of geometrical forms used in decorations. In particular, it is worth seeing that the ribbon ornaments from the Neolithic period all had 7 one-dimensional crystallographic groups on the surface.... However, we cannot be certain that some kind of geometrical reflection was followed. (p. 23)

Pre-historical art might be proof of humans' interest in the shape of figures and in the symmetry and the rhythm of linear and 2-D arrangements.

The Pythagoreans, in their philosophy, adopted the idea that the world was orderly through numbers and the relationships occurring between the numbers. More than the individual figures and separate numbers, the Pythagoreans were interested in the relationships between them. In particular, they chose a visual representation of numbers and arithmetic relationships; they formed "evidence pebbles" for figural numbers. The series configuration formed there suggestively indicated the existence of a common idea involving all the elements of the series and forced the analysis of described compounds. They felt this strange harmony and the power of exploring this harmony (Sękowski 1996, p. 23).

Order is present not only in the universe, but also in art, music, and architecture. The recognition or realisation of isometric drawings has been strictly connected with the artistic tradition. As Weyl (1952) wrote, in the course of time, isometries occurred at the beginning as practical knowledge for art and architecture; they later became mathematical objects when the interest of mathematicians turned towards the "theory of transformations" and the study of "invariants." The fundamental idea of Klein's program (1872) was that not only Euclidean geometry but also other geometries exist, each of them associated with a group of transformations. In particular, the study of the "group of movements" and the connected invariants leads to the "geometry of isometries." The word *isometry* is composed from ισος, "the same," and μέτρον, "measure": it contains the concept of a transformation that conserves the distance. There are isometries that also conserve the orientations of the figures (direct isometries), i.e., translations; other isometries do not conserve the orientations (inverse isometries), i.e., the mirror symmetries.

All peoples in all the ages have used patterns in their artistic expressions. A *pattern* could be a horizontal sequence of decorative drawings or a bi-dimensional tessellation made using a regular repetition of a geometrical motif and following rules. The observation of the following horizontal patterns, obtained using some of the letters of the alphabet, shows that some of them present regularity (*a, b, c, e*) and others do not.

We can observe that all the patterns in Table 1 are obtained using the letter *p* and its "transformations": to pass from the letter *p* to the letters *q, b*, or *d*, we use reflections, rotations, or glide reflections, i.e., isometries. Moreover, by using isometries, we find that all the horizontal regular patterns can be organized in a limited way: there are only seven kinds. Here we present seven patterns obtained using a shape that looks like the letter *L* and all its possible isometric transformations. Starting from this letter, in each case we created a regular motif and, with its systematic repetition, a pattern (Fig. 7).

Table 1 Examples of some horizontal patterns

a	p d p d p d p d p d	d	p d d b p d b d p b
b	d b d b d b d b d b d b	e	d b d b d b d b d b d b
c	p q p q p q p q p q b d b d b d b d b d	f	q p p qp pq q p q p p

Fig. 7 Seven one-dimensional groups of isometries on the surface

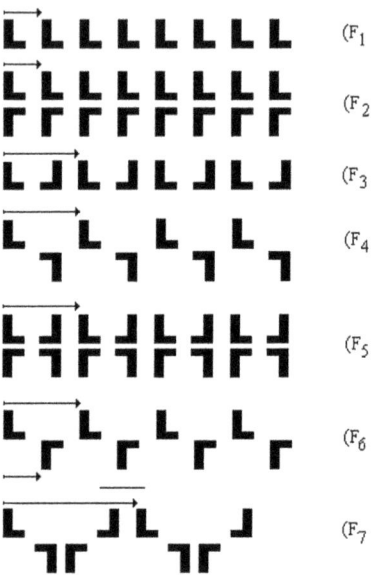

Recently, Devlin (1996) re-evaluated the role of patterns in mathematics. In fact, they are very rich from the conceptual point of view, so they are important in mathematical education. Activities with patterns involve many abilities: observation, ordering, copying, iteration, use of translations, rotations, symmetries (isometries), control of distances, and direction. Work with patterns can be of two different kinds: construction of a sequence following an assigned law or individuation of the rule followed in a sequence. This requires the coordination of sight, manipulation, manual skills, and possibly language and argumentation. Verbalisation promotes the transition from a static to a dynamic approach to the isometries and, in particular, the conception of the isometries as mathematical objects (Marchini and Vighi 2011).

3.5 Reason for Research Related to Geometrical Patterns

In a discussion about relevant research in this field, Waters (2004) states that exhaustive research has revealed that mathematical patterning has generally not been studied as a topic area in itself but has been part of broader studies (p. 566).

Moreover, the relationship between children's ability to make patterns and their future mathematical development is not so clear (Clarke et al. 2006). This is evident in the following questions posed by Economopoulos (1998):

> What should be looked for when young children work with patterns? What kind of markers are there to identify development? What is really known about young children's engagement with pattern and the thinking that is being displayed? Are opportunities being provided for reasoning and justification that take the activity beyond "what comes next"? (Economopoulos 1998, p. 232)

All the above have led us to perform research with a focus on patterns. We were particularly interested in studying how patterning can stimulate and shape young children's intuitions about geometrical transformations.

We can look at children's actions with geometrical regularities in two different situations. One is when children create their own regular compositions: patterns, buildings, and mosaics. Another is when they see an existing pattern and are asked to continue it.

3.6 Results of Research in Geometrical Patterning

3.6.1 Is It Obvious How to Continue?

"To see that" does not necessarily mean "to know how" (Swoboda and Tatsis 2010; Swoboda 2013): a child may detect a geometric regularity but may not be able to create it alone (create = continue).

Although the perception of regularities in symmetrical arrangements is in the realm of even small children, it is not true that it is an innate ability. The youngest children can have difficulty in continuing a rhythm or are not able to read a regularity at all. Sometimes they finish a task by drawing anything. They act on the social level; when asked to draw they draw "something."

Example 1 (Krystian, 4 years old; Michał, 4 years old)

Teacher: I have prepared a pattern for you. Do you know how to continue the drawing?
Krystian: A little person.
Michał: A house.
T: How do you know it?
K: Because I know everything (Figs. 8 and 9).

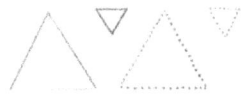

Fig. 8 Michał's work

Fig. 9 Krystian's work

In this situation, the children took the space on the paper as a place for their own free creativity. They did not treat this task as a continuity pattern with regularities and relations between figures. They ignored the information given by the teacher that they had "to continue" what was already begun. Those children were not yet ready to perceive (and look for) regularities.

In other situations, children may create their own rule according to which they continue the work.

It is also possible that children quickly create their own rule that is incompatible with the idea of the suggested rhythm. This is shown in Example 2.

Example 2 (Oliwka, 4.2 years old)

T1: Oliwka, listen and look. What figure will be next? Try to draw it. Triangle, circle, triangle, circle, triangle…
O1: Triangle, I know what's going on. [She draws the drawing shown in Fig. 10a].
T2: Listen and look. Circle, circle, square, circle, circle…
O2: Circle. [She draws the drawing shown in Fig. 10b].
T3: Should we do the next task together or do you want to draw alone?
O3: No, no, please read it.
T4: So, triangle, rectangle, triangle, rectangle…
O4: Rectangle, as I have shown. [She draws the drawing shown in Fig. 10c].
T5: Now the last rhythm. Rectangle, rectangle, triangle, triangle, rectangle, rectangle, triangle, triangle.
O5: Triangle… It is a pity that it is a triangle, because I do not know how to draw it. [She draws the drawing shown in Fig. 10d].

Olikwa developed a kind of strategy to deal with this task. Just after the first experience, she developed and set the rule according to which she continued the work. She finished each example by repeating the last word spoken by the teacher and by drawing an appropriate shape. She understood that the task was about the repetition of figures but still did not see the proper rhythm.

(a)   (b)   (c)   (d)

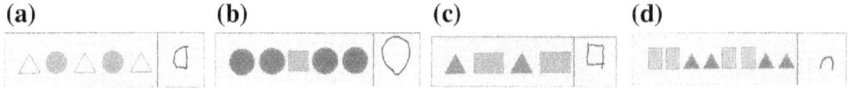

Fig. 10 Oliwka's responses to continuing pattern problems presented her by the teacher on the sheet of paper

3.6.2   Perceiving Regularity

As opposed to what was shown above, other research results involving children's reasoning have shown that they are able to see general rules and feel that some statements are very general. They are even able to express the generality of rules using their particular means of expression in the following ways:

a. **Rhythm realised by the replacement of two (or more) different elements simultaneously placed**

In geometrical patterns, different kinds of regularities exist. The most obvious are patterns where the rhythm is created by putting two different elements in a sequence: A, B, A, B, etc. This type of rhythm can be realised by many children (Fig. 11).

The rhythm, however, may became complex, and regularities may lie at a perceived eligible distance among figures or in a suitable arrangement. Mutual relations are connected with some geometrical relations and transformations, such as parallelism, perpendicularity, translation, rotation, axis symmetry, and similarity. Those relations exist on the intuitive level only. Rhythm existing in a pattern can reasonably direct a child's mind to very abstract concepts, such as an infinite straight line or the measurement of distance between figures. Students can understand the principles valid in drawing patterns. They can grasp general rules and are able to work consistently according to those rules. Their work shows that there is not merely visual copying of the motif given. Children are able to concentrate on at least one geometrical phenomenon and to work consistently according to this phenomenon. Other situations may also occur—at the beginning, children may not be able to co-relate all the conditions in the pattern, but during their work they can improve their ability (Fig. 12).

Sometimes technical problems can occur; a child's manual efficiency may be too low and the child may not not able to reproduce some complicated shapes. The child is then focused on a chosen part of the task or on selected features of the observed regularities (Fig. 13).

Fig. 11   Linear sequence continued by a 6 year-old girl

Fig. 12   Struggles with the rules in a linear sequence in the work of a 6 year-old boy

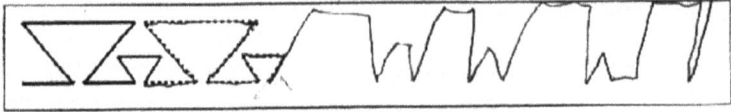

Fig. 13 Observed relation represented by a 6 year-old boy with manual problems

Fig. 14 Focusing on the "big-small" relation, represented by a 6 year-old boy

- **Realisation of some specific relation between figures**. A specific relation between figures that have the same shape

The size relation (scale) is a very hard problem at this educational level, but grasping an intuitive relation is possible. A deep analysis into recognising the rules can lead to the separation of a basic motif and its repetition (Fig. 14).

- **A straight line**

Children draw figures "equally," often obeying that rule in a conscious way. The line exists not only as the bottom and upper limitation, but also in the middle of the pattern. In Fig. 15, the bottom line goes beyond the frame into infinity. This boy had great difficulties in keeping all the conditions existing in this pattern at the same time, but the idea of an unlimited straight line was the most important for him.

- **Parallelism**. Sometimes the parallelism of sides is more important than the shape of the particular element (Fig. 16).

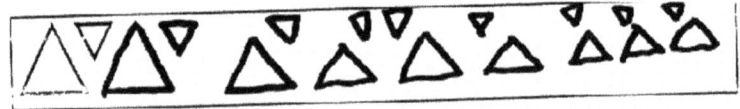

Fig. 15 Unlimited straight line created by the bottom sides of triangles; the work of a 6 year-old boy

Fig. 16 Parallelism of the sides of successive triangles

b. **Abstract thinking, discovering structure**

One can assume that patterns, having such a complex geometrical structure, are good tools to introduce abstraction. Children are able to perceive various structures in the pattern. Some children's work shows that children focus on selected geometrical aspects. There are some examples below, with descriptions (Figs. 17, 18, 19, 20 and 21).

Fig. 17 Pattern consisting of various triangles, freely arranged

Fig. 18 Pattern consisting of two separate strips of triangles: one band at the top and one along the bottom

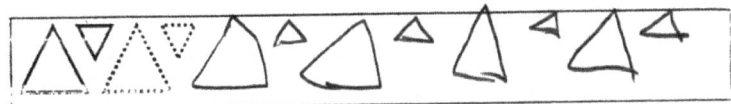

Fig. 19 Pattern consisting of consecutive triangles: large and small ones, with the small triangles placed at the top

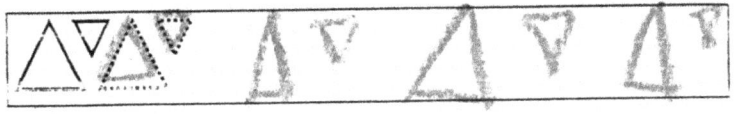

Fig. 20 Pattern with a basic motif made up of a large triangle and a small triangle

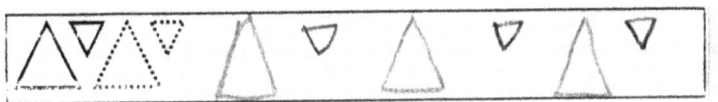

Fig. 21 Pattern consisting of consecutive triangles, large and small, with the small triangles placed upside-down on the top with an equal distance between them

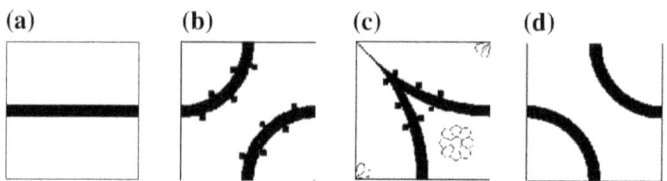

Fig. 22 Kuřina's tiles. **a** Straight, **b** Branch, **c** Flowery, and **d** Swallows

3.6.3 Arrangements of Surfaces

The following four different kinds of tiles were created by Kuřina (1995):

From now on, we will refer to the tiles using the names invented by our pupils, as shown in the caption for Fig. 22. These names in themselves show a naturalistic interpretation of the tiles, not a geometrical one. The fourth, Swallows, is dynamic, whereas the other three are static.

The drawings placed on the tiles were carefully studied: tiles *a*, *b* and *d* present two orthogonal axes of symmetry, while tile *c* has only one axis of symmetry. Furthermore, the drawings offer the option of connecting the tiles to obtain a continuous path (involving the mathematical concept of continuity).

a. **Feeble and rigid structures**

Swoboda (2005, 2006) conducted research based on Van Hiele's theory (Van Hiele 1986) that focused on children's interpretation of protocols. The same activity was implemented in Italy by Marchini and Vighi (2007). The three authors cooperated and developed the research in a common work (Marchini et al. 2009).

Van Hiele's (1986) theory of levels, which studies geometrical thinking and understanding, divides the educational processes in geometry into different levels. In particular, the analysis of the three initial levels shows a very important aspect: manipulation, which occurs at every level. Van Hiele also distinguishes between rigid or feeble structures.

> In his opinion, feeble structures are worth noticing; they fill out the majority of our everyday life. They come from a non-verbal, intuitive way of thinking, but mathematical thinking is not superior to the intuitive one. Feeble structures could be the beginning of knowledge on a higher level, which we can act with rigid structures or possibly still feeble ones. (Marchini et al. 2009, p. 63)

Manipulation is promoted using the following task: Create from these tiles as beautiful a floor as possible. The request is intentionally ambiguous, and the children are completely free to choose which and how many tiles to use and where and how to place them in order to obtain their own floor. The "floor" consists of an A4 blank sheet of paper (21 cm × 29.7 cm), and the side of each tile measures 2.5 cm. Obviously, the choice of these measures is not a random one, but is motivated by different reasons.

An analysis of protocols allows an initial classification to be made based on the criteria used by the pupils in the construction of the floor. We distinguish the following kinds of criteria:

(1) Random: Pupils glue the tiles as they pick them up at random, without observing the drawings on them.
(2) Taking the drawing on the tile into account: On the straight tile, the line is parallel to the edge, so the children tend to glue the tile with the line horizontally or vertically. Other tiles do not have a preferred direction, although tiles tend to be placed with their sides parallel to the edges of the paper as far as this is possible.
(3) Influenced by and based on neighbouring tiles: Construction of a route, translation, or symmetry; construction of a flower in the case of Flowery.
(4) Regular: An iterative and regular tessellation.
(5) Progressive conquest of regularity: Initially pupils glue tiles at random and subsequently choose regular tessellation.
(6) Project: Pupils first "see" a mental representation and then proceed to the concrete manipulation. Sometimes they are unable to specify their mental image (feeble structure).

The drawings on the tiles are such that when the same type of tile is placed next to another of the same type, the lines fit together perfectly. This feature leads pupils to imagine specific things. For example, Swallows or Straight might give the idea of a road; Branch might give the idea of a scene from nature, such as a thorny lawn; and Flowery might give the idea of a garden. However, the pupils' imagination is even more fertile than this: they "see," for example, a chick and a cat in the arrangements of tiles shown in Fig. 23.

More mature pupils create a greater number of basic motives (Budden 1972). The use of the flowery tile increases in the second grade.

We obtained protocols showing feeble or rigid structures (Fig. 24).

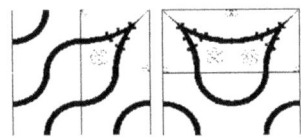

Fig. 23 Chick and cat

(a)	(b)	(c)	(d)	(e)

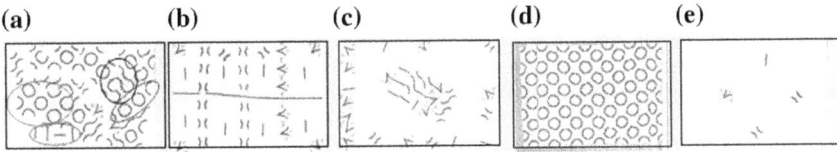

Fig. 24 *Black* and *white* protocols

Protocols 24a and 24c were produced in kindergarten, a first grader made protocol 24d, and the authors of the remaining protocols were second grade pupils. We have superimposed some ovals on the reproduction of the original protocol in Fig. 24a in order to draw the reader's attention to feeble structures. The other protocols show the presence of rigid structures.

b. Different concepts of space

The activity of paving allows children's concept of the space to be investigated. In geometry, space is unlimited. In fact, in children's activities space is reduced to the limited surface of the sheet of paper where they are obliged to carry out the task. But the way children utilized the paper gave answers to the research question: How does a child fill the floor and, as a consequence, does the child conceive the space represented by the sheet of paper to be limited or unlimited? Two opposite approaches are possible: by creating the floor, the child clearly remains on the interior of the paper, devising a "picture with a white frame" (limited space), or the perception of the regularity pushes the child to continue the floor even if some of the tiles go outside the boundaries of the sheet of paper (unlimited space).

"Greek thought... tried to escape from the unlimited, considered as a form of imperfection. For Aristotle there is no space above the sky of fixed stars" (Speranza 1997). This is the same concept that led some learners in our activity to not attach tiles that would go over the edge of the sheet of paper. Sometimes children clearly limited the space by creating a "frame" (Fig. 25).

Other learners conceived space as being unlimited and thus had an "in act" conception of infinity (Marchini 2004). What provoked them to cross the edges was the idea of regularity (Fig. 26).

Another distinction is between "intra-figural space" and "inter-figural space" (Vighi and Marchini 2014). The first is denoted by locutions as *horizontal, vertical*, and *crooked*, referring to the sheet of paper's sides, which constitute a reference system. The second is revealed through the child's statement regarding distance among the tiles or the presence of space between them (Fig. 27).

A consequence of an intra-figural space idea is that the space is anisotropic, i.e., there are two privileged directions, horizontal and vertical, while an inter-figural space is isotropic, i.e., without distinction among directions.

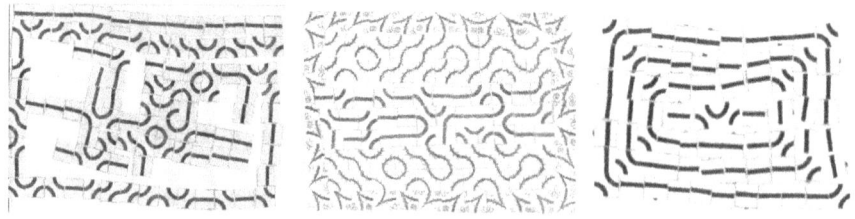

Fig. 25 Limited concept of space represented by the sheet of paper (*left* Boy, 4 years old, frame as element of composition; *centre* girl, 7 years old; *right* boy, 6 years old)

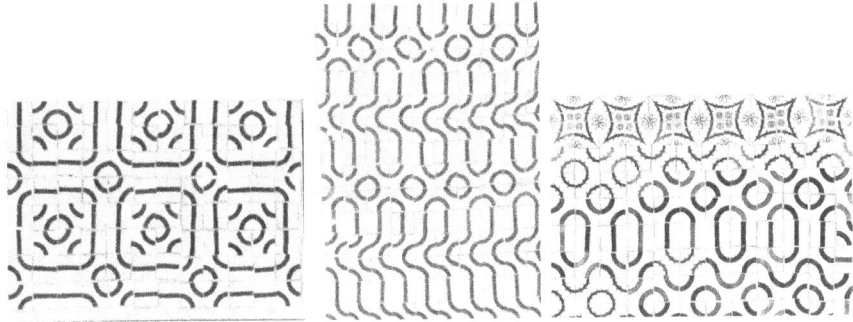

Fig. 26 Unlimited concept of the space represented by the sheet of paper (*left* girl, 6 years old; *centre* boy, 5.5 years old; *right* boy, 5 years old)

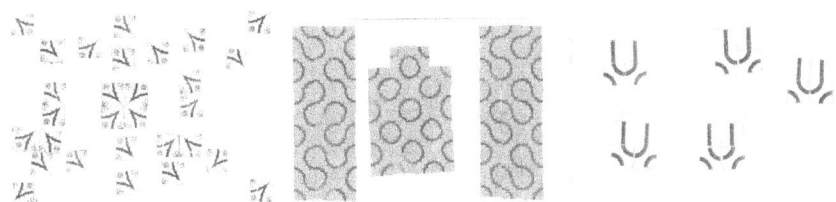

Fig. 27 Examples of inter-figural space (*left* boy, 6 years old; *centre* girl, 7 years old; *right* boy, 7 years old)

c. **Connection and continuity**

The drawings on the tiles allowed paths to be created by gluing the tiles next to each other with suitable orientations. The use of connections appeared more frequently with tiles of the same kind. The presence of connections was greater in male protocols. Moreover, it increased with the age of the children. The continuity concept, achieved by the connections, was present in a large majority of protocols. Figure 28 shows the possible basic motives created by four tiles of the same kind.

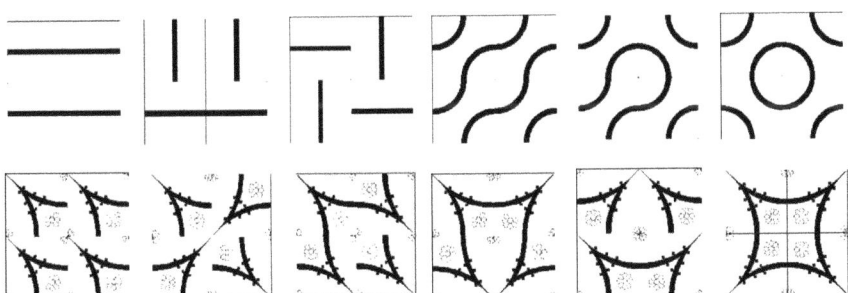

Fig. 28 Examples of tile arrangements

d. Coloured protocols

A second phase of the Italian experiment was based on the use of colour: children received photocopies of their "black and white" protocols with the task of colouring it, also assigning a title to the "coloured floor." The introduction of these two new variables, colour and title, allowed semantic aspects to be analysed. Colour is a kind of language that reveals the criteria the learners base their design on.

Colour provided new information about the pupil's aims and significantly changed the possible interpretation of protocols: in some cases, the colour respected the rigid or feeble structure of the black and white protocol (Figs. 29a, c), but we also had the case that colour transformed a feeble black and white structure (Fig. 30a) into a more rigid one (Fig. 30b). Figure 30a, b show the superposition of a vertical-horizontal structure given by colouring onto a diagonal one present in the black and white protocol. The tiles were sometimes employed for figurative/decorative aims, as if each tile were the mark of a pencil. An example is in Fig. 30c, "The boy looks at the sun."

On other occasions, the colouring and title revealed the child's aims (Fig. 31a, b: "A tree and the rain").

Colour gave opportunities for revealing different interpretations of geometrical aspects: it could show where the child focused attention. The drawings in Fig. 32 used tiles of the same type, but children "found" space in different parts of the tile, highlighting it with colour.

It is evident that there were different approaches in the four drawings of Fig. 32. In Fig. 32a, the space of the tile is on the "concave side" of the lines, making a non-connected space. The author of Fig. 32b was interested in the lines themselves

(a)   (b)   (c)   (d)

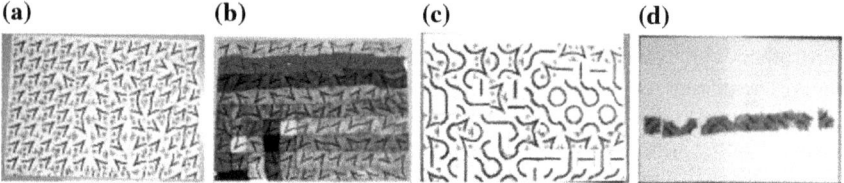

Fig. 29 Examples of coloured protocols

(a)   (b)   (c)

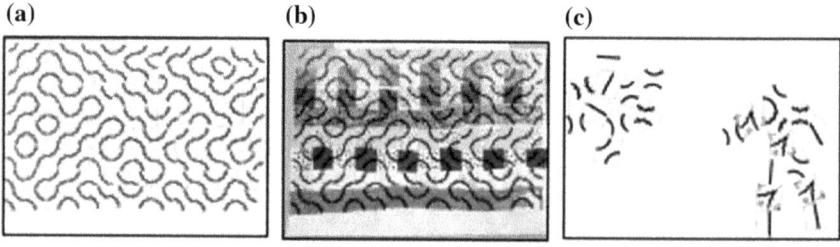

Fig. 30 Changing the structure with colours

(a)   **(b)**

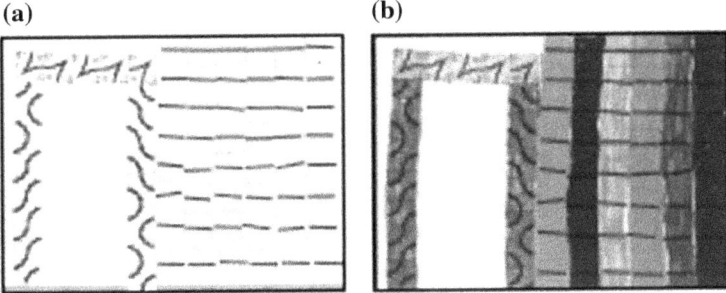

Fig. 31 a, b A tree and the rain

(a)   **(b)**   **(c)**   **(d)**

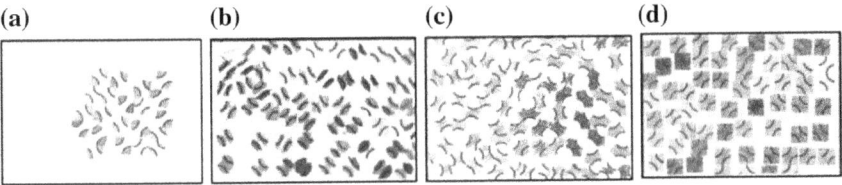

Fig. 32 Colours reveal different concepts of space

and the colour highlights them. The space between the "convex parts" of the lines on the tile attracted the pupil who created Fig. 32c. For the author of Fig. 32d, the space was the square tile, each of which has been coloured separately.

The geometrical tile structure could limit the degrees of freedom for children's expression. Furthermore, the mind activity required to construct and colour the drawing is, in our opinion, a suitable task for the spatial-geometrical activity necessary to prepare the next more formal treatment of geometry.

3.6.4   Geometrical Relations

In many situations, children are guided by one idea with possibly only a few distractions. Some relations can be described using geometrical relations. Rhythms and regularities realized by a child on the visual level can be described by the language of transformation or by the language of the relation between figures, but these mathematical definitions are not the conditions that lead a child's work. The core of the structure at the visual level is different.

a. **Translation**

The search for intuitions while translating children's works consisted of finding works in which regular repetition of a certain motive occurred. Such regular repetition had to keep the parallel arrangement of segments and lines. The second

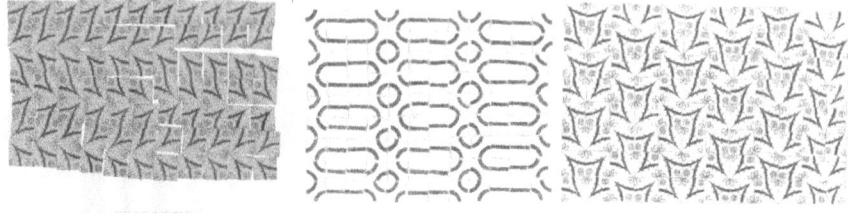

Fig. 33 Examples of translation (*left* boy, 4 years old; *centre* girl, 6 years old; *right* girl, 8 years old)

important condition was the preservation of equal distance between the elements. Some examples are shown in Fig. 33.

b. **Axial symmetry**

There were also situations where a child created a symmetrical, limited construction. Its creation proceeded floor by floor. The whole work then had only one symmetrical axis (Fig. 34).

Table 2 presents a quantitative analysis of the presence of isometries in the Italian pupils' protocols.

Table 2 reveals that the rate of application of the more complex isometries grows with the pupils' age, even if the children in the sample did not receive any formal geometry teaching. We can interpret this data by observing that the enhancement of manipulation ability obtained by school teaching is probably also of benefit for improving geometrical and spatial intuitions. Marchini et al. (2008) treats these and other interesting geometrical features discovered in protocols in depth. It is possible to observe a low number of symmetries, even if the drawings on the tiles could suggest their use. This supports the ideas that the "metaphor of equilibrium" (Nùñez et al. 1999) does not influence the pupils' performance and that symmetry is not an embodied cognition but mathematical knowledge that must be constructed by learning (Swoboda 2007; Bulf 2010; Bulf et al. 2013; Bulf et al. 2014a; Bulf et al. 2014b).

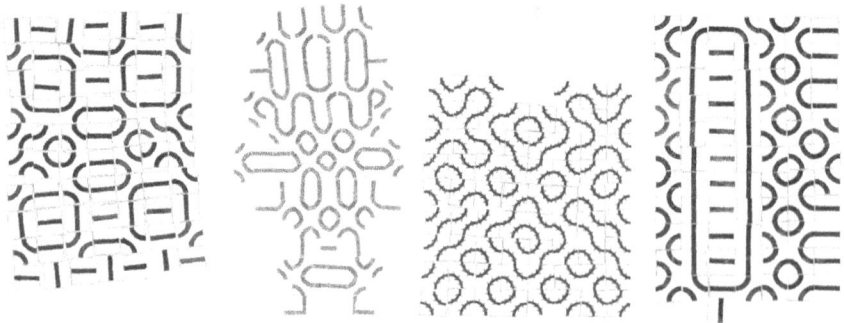

Fig. 34 Examples of axial symmetry (*left* boy, 7 years old; *centre left* girl, 6 years old; *centre right* boy, 6 years old; *right* boy, 6 years old)

Table 2 Presence of isometries in children's work

Age (years)	Local relation (in %)			Global relation (in %)		
	Translation	Axial symmetry	Rotation	Translation	Axial symmetry	Rotation
4	43	0	0	0	0	0
5	30	12	8	4	1	1
6	28	18	9	5	4	1
7–8	34	26	13	13	6	2

3.6.5 Quantitative Analysis of the Protocols

The quantitative analysis involved only the protocols produced during one school year in the research in Italy (Marchini and Vighi 2007). The sample consisted of 212 pupils (97 in last year of kindergarten, 68 in first year of primary school, 47 in second year of primary school; 122 male and 90 female) working individually in a classroom environment. The protocols were analysed using different indices.

a. **Covering index**

The first analysis is based on counting the numbers and the types of tile the children used in their protocol production. Possible influences on the number of tiles included level of attention, manual coordination, commitment to the task, and interest in the floor's production and design.

The A4 sized paper (21 cm × 29.7 cm) and the size of each tile (2.5 cm) allowed exactly 88 tiles to be used per page if they were placed contiguously and did not overlap the edge the paper. Therefore, we considered 88 to be the *theoretical covering index* (in the following table, it is assumed to be 1). Only 11 pupils (5.19 %) used exactly this number of tiles. Thirty-two children (15.09 %) chose to extend the paving beyond the sheet edges, using 96 tiles (constructing a floor 8 × 12, they covered a hypothetical sheet of 20 cm × 30 cm). Fifty-eight protocols (27.35 %) presented a number of tiles between 88 and 96 tiles and 35 protocols (16.51 %) used more than 96 tiles. The high numbers of tiles used by the majority of pupils clearly shows that the activity was motivating for them.

Table 3 shows the average covering index values. The data clearly shows differences among ages and between genders. It is possible to observe that the index increases with the years of schooling. The presence of high scores for particular classes may indicate pupils' possible previous independent experience of a similar activity or different teaching methods. In conclusion, the results confirm that the tools employed were useful for investigating different ways of thinking.

b. **Diversity index**

Another quantitative analysis is connected with the count of each kind of tile used in the protocols. This enables use of Shannon's diversity index, originally used in biology as a measure of entropy (Shannon and Weaver 1949), but borrowed from

Table 3 Average Covering Index

School	No. of pupils	Average Covering Index	Male Average Covering Index	Female Average Covering Index
Kindergarten	97	53.5	47.9	59.8
First grade primary school	68	77.1	79.1	74.0
Second grade primary school	47	86.5	87.2	85.4

the mathematical theory of communication. This index varies between *0* (every tile in the protocol of the same kind) and *2* (equal number of tiles of each kind in the protocol). In our research, the diversity index was calculated as follows. Let N be the total number of tiles used (in a protocol, in all protocols of a class, in a school, and in a type of school, respectively). Let n_i ($i = 1, 2, 3, 4$) be the relative number of Flowery tiles (1), Branch tiles (2), Straight tiles (3), and Swallows tiles (4) used in the protocols (of the class, school, etc.). First, we calculated the rate of presence of each tile, then the diversity index D, using the Shannon and Weaver (1949) formula:

$$D = -\sum_{i=1}^{4} \left(\frac{n_i}{N}\right) \cdot \log_2 \left(\frac{n_i}{N}\right)$$

Table 4 shows that the diversity index is fairly constant, with a small decrease corresponding to added years of schooling. In biology, an index near 2 shows good "ecological health." In our experiment, the high values of this index can be interpreted as a measure of the pupil's attitudes or as the respect of practices introduced by the teacher.

The data in Table 4 shows the number of tiles used by males or females in paving their floors.

The Table 5 shows a clear difference between males and females. The tiles seem have a gender connotation. The way in which girls and boys use the Flowery tiles is particularly striking. The "monopolization" of the Flowery tiles by girls somewhat lowers the diversity index. These facts are therefore connected. Figure 35 represents both indices with graphics.

Table 4 Average diversity index

School	No. pupils	Average covering index	Male average covering index	Female average covering index
Kindergarten	97	53.5	47.9	59.8
First grade	68	77.1	79.1	74.0
Second grade	47	86.5	87.2	85.4

Table 5 Rates of tile use

	Flowery rate of use	Branch rate of use	Straight rate of use	Swallows rate of use	Total no. of tiles
Sample	24.45	13.61	13.75	17.73	14,740
Sample males	15.77	16.18	14.12	**21.91**	8294
Sample females	**36.21**	10.12	13.23	12.06	6446
Kindergarten	28.91	15.50	15.76	22.85	9548
K. males	18.17	17.27	17.15	**29.82**	5851
K. females	**46.25**	12.66	13.50	11.61	3697
First grade	22.79	10.88	14.04	29.41	5245
First grade males	14.69	14.36	12.55	**37.50**	3332
First grade females	**35.88**	5.27	16.46	16.35	1923
Second grade primary	37.77	16.83	18.06	13.85	4066
Second grade males	23.21	21.48	**23.83**	18.69	2529
Second grade females	**61.22**	9.33	8.78	6.06	1537

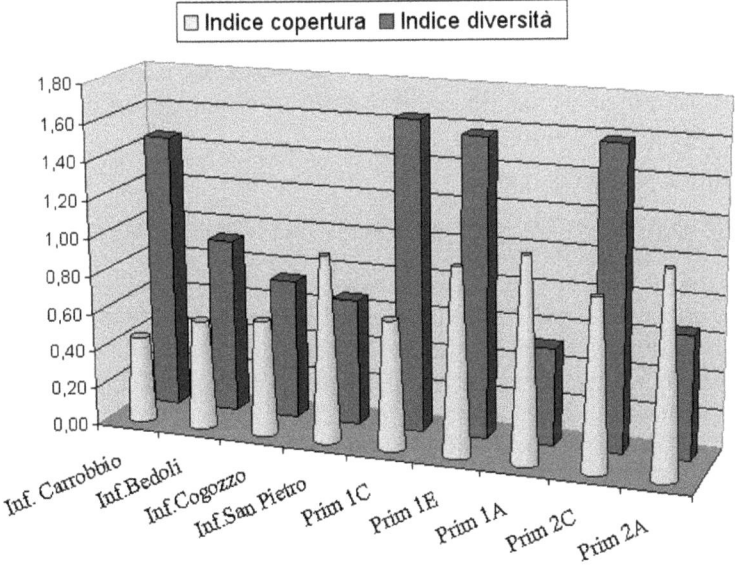

Fig. 35 Average covering index and diversity index

Fig. 36 A "path" with "stones"

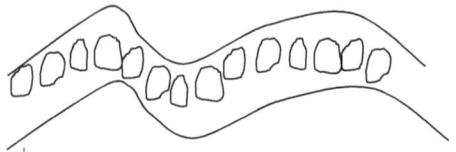

3.6.6 Building Vocabulary, Development of Language

Many geometric activities can take place without words and without symbolism. Visual language is very rich in content so it is possible to use it at very early educational levels. The act of constructing patterns and making tessellation requires a long sequence of elementary acts: observation, ordering, copying, and repeating. Swoboda (2005) showed that drawing a pattern is not merely perceptual copying, but is a deep thinking process, which involves the body and gestures (Marchini and Vighi 2005). Of the domains of knowledge where children must enter, geometry is the one that needs the fullest cognitive activity, as it requires gestures, language, and seeing. It requires the child to construct, reason, and see; each activity is indissoluble from the others (Duval 2005). Arzarello (2004) emphasized the role of body movements and gestures in learning. Gestural expressiveness can be considered a sort of language useful to understand pupils' thoughts, taking into account the poor language competencies of children of those ages. Geometry as rhythms and patterns gives a chance to code and de-code rules and formulas at a very low level. The passage from perception through geometrical symbolic representation to verbal mathematical description is very long and often strange for children. Engagement in activities creates a good opportunity to create a space for discussion. Children start to transform these relations into words. By talking, these relations gain the status of existence. They emerge gradually from experience and start to be facts related to the mathematical world.

a. **Creation of argumentation**

Example 3 (Work in a small group)

Children made a paper "path," drew "stones," and painted them in three different colours, making patterns. The next day, the groups exchanged "paths." The teacher covered one stone with a sheet of paper. Children were then asked to say what colour this stone was and give the reason why their opinion was correct.[1] Children could use general, abstract argumentation to see "a general in particular" (Fig. 36)

Ola (5), Wiktoria (5), and Michał (4)

T: Who knows which colour the covered stone is?
Michał: White.
T: And why do you think that it is white, Michał?

---

[1]The Italian didactician, B. D'Amore is the author of this task.

M: Because it is.

T: And Ola, what do you think?

Ola: Green.

T: Why do you think that it is green?

O: Because here is red [she shows a stone lying before the covered one], and here, after a red stone there is a green one [she shows a group of stones lying earlier, but not directly before the group with the covered stone].

T: And what do you say, Wiktoria?

Wiktoria: The same.

T: Why do you think that it is green?

W: Because it is here and there [points at stones lying on both sides of the covered stone], and here in the middle there is a green one.

The two 5-year-old children noticed a rule, but their arguments were different. The sense of Ola's comments was as follows: "If the stone before the covered stone is a red stone, then the covered stone has to be green because after the red one there is always a green one." Wiktoria's stated that in the whole pattern there was a basic motif of *red-green-blue*, so if the covered stone had a red stone on the left side and a blue one on the right, she could be sure that in the middle was a green one.

The arguments given by Ola and Wiktoria are examples of two different approaches to the problem given to the children. Ola gave her opinion on the basis of directly (empirically) perceived relations between the other group of stones and one particular group with the covered element. The perceived group existed physically. The valid principle was: I see an arrangement in a group of stones => I see an arrangement in a group of stones only in a part. Wiktoria's reasoning presented was more general. She understood the general rule in the pattern. The valid principle was: arrangement in the basic motif ⇔ arrangement in each concrete motif.

b. Geometrical transformations as an instrument of reading

Research implemented in Grade 3 by Marchini and Vighi (2011) proposed an "instrumental approach" inspired by the "cognitive ergonomics" of Verillon and Rabardel (1995), in which artefact is the main tool of semiotic mediation (Bartolini Bussi and Mariotti 2008). The main idea is that an innovative introduction of isometries can play a relevant role in integrating the traditional teaching practice with the geometry of transformations. The research questions were: Are isometries a suitable topic for grade 3 pupils? and Does learning isometries affect the "standard" Italian school geometry?

The artefacts employed were the following protocols (Fig. 37), previously realized by children of another school, using the same tiles shown in Fig. 22. The criteria of the choice were based on the presence of particular isometries: translation in 2A15; translation and symmetry in 1A16; and translation, symmetry, and rotation in 1A17 and 2A16. In particular, the protocol 2A16 was chosen on purpose since it presented a mistake in order to make the presence of a rule in the construction of the protocol evident using its violation. The protocols 1A17 and 2A16 presented

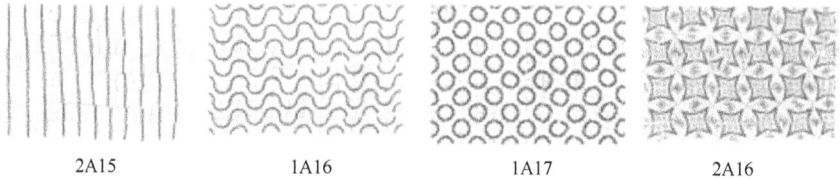

| 2A15 | 1A16 | 1A17 | 2A16 |

Fig. 37 Treatment protocols presented to the pupils

the same structure but obtained with different tiles: it could hide the fact that the construction rule was the same.

Pupils discussed the documents of Fig. 37 in groups, recorded the worthy aspects, assessed protocol, and orally presented their conclusions. Afterwards, for each protocol the researcher proposed an "institutionalisation" activity based on drawing four squares on the black board, as shown in Fig. 32, schematically reproducing the four consecutive tiles as they appear in the upper left corner of the document and asking whether there was a "tie-in" between two consecutive tiles. The aim was to verify the recognition of the rotated, translated, and reflected figures (MPI 2007). Referring to "floor" 2A15, some pupils suggested the word *trasloco* (move). For 1A16, the word *specchio* (mirror) came out without difficulty when looking at the disposition of horizontal tiles. Fortunately, in the initial letters of the words *trasloco* and *translation* are the same in Italian and English, which is also the same for *specchio* and *symmetry*. This allowed the names of two different isometries to be translation (T) and symmetry (S). Pupils understood that T works in 2A15, both from left to right and from up to down, while in 1A16 S acts from left to right while T works going up to down. Moreover, they found that 1A17 was the result of applying S both horizontally and vertically. Analysing the 2A16 protocol, they discovered that its author made a "mistake" and they suggested that it was in a wrongly rotated tile. This observation furnished the opportunity to introduce the word *rotation* (R). Secondly, with the aim of considering isometries as mathematical objects, the researcher presented a card game based on the use of a game board (Fig. 38), an array of two times two squares, and three kinds of "playing cards" with the letters T, S, and R in a box. The rule was to draw four times, with restitution, one card at a time and to write the letter of that card in the empty spaces, in this conventional order: top-left-right-bottom. Finally, a tile was placed (or drawn) in the top left board square. Pupils copied this in their exercise books. For example, with a suitable choice of the tile and its placement we get T, T, T, T for

Fig. 38 The game board

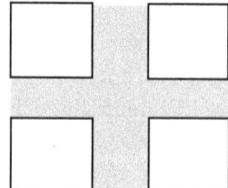

2A15; *S, T, T, S* for 1A16; and *S, S, S, S* for 1A17 or 2A16. The card game with isometries could be used in every school environment to raise isometries, both as procedures and as mathematical objects.

Research into the first aim had a large number of corroborations during the treatment (Marchini and Vighi 2011) and there was also permanence of the taught concepts one year later, e.g., by words or drawings mentioning isometries. Pupils very often traced lines onto the figures showing how it was possible to decompose the shape in a fixed part and in parts that correspond under a local isometry. These lines reveal a sort of mereological decomposition (Duval 2006). Therefore, the second aim of our research was achieved.

3.6.7 Connections with Other Geometrical Aspects

The other aspect of the same issue suggests not only paying attention to teaching basic notions, but also to building the intuition needed for further creation of geometrical notions that will appear later.

a. **From arrays to the concept of area**

While working with tiles to construct a floor, there is a need to cover a plane without holes and without superimpositions of tiles. This means that it is necessary to put the tiles near each other covering the plane. A child who leaves "space" between the sides of the tiles still does not have the idea of covering. The same problem appears when a pupil superimposes some tiles, maybe trying to avoid leaving the sheet of paper. Moreover, the edges of the sheet suggest the organization of the array in horizontal and vertical directions. Rożek and Urbanska (1998) studied row-column arrangements in depth. They affirmed that the concept of bi-dimensional structure is not natural but must be constructed: "The children have a different awareness of the rows and columns arrangement. Some of them prefer rows, some of them columns. It appears that it was difficult to see both rows and columns, especially for young children" (Rożek and Urbanska 1998, p. 304).

The activity of paving a floor can promote the acquisition and mastery of the row-column arrangements often employed in mathematics (Vighi 2006). For instance, it is the first and important step towards the idea of bi-dimensional distribution and the concept of array that underlies the measurement of areas (and multiplication too).

b. **From tessellations to the real world**

The activity presented could help the observation of some human artefacts, such as geometrical decorations, carpets, and textiles, in terms of isometries. In our experiment (Marchini and Vighi 2009a, b), Escher's drawings were employed as artefacts to verify whether the practice with isometries on simple drawings could be transposed to complex non-standard shapes. The aim was to attach attractive and effective aspects to transformations. For each drawing presented, the first question was "What can you see in this drawing?" The task then became individualised with letters or colours, with many figures obtained from a starting figure by a suitable isometry. Children showed no difficulties with translations, and the recognition of rotations was frequent. Frequently, the spontaneous use of central symmetries appeared, while the recognition of axial symmetries was the most difficult.

4   Summary and Looking Ahead

In this Topical Survey ICME-13 you can find:

- Main features of early geometrical thinking;
- Identifying and creating shapes to construct the concept of geometrical figures;
- Different kinds of isometries and difficulties in their employment;
- From ability in making patterns to mathematical reasoning development; and
- Activities about regularity: examples, analysis, and proposals.

Results of research related to secondary school pupils' conceptions of geometrical objects and relationships show that this knowledge is not well established. One of the reasons is the very poor recognition of the way geometric knowledge develops in the early educational stages.

Fortunately, research related to early geometry has started to become increasingly important. This has happened in parallel with the conscious return of the teaching of early geometry in schools. Without denying the relevance of understanding numbers and operations for young students, it is essential to highlight the increasing role of geometric reasoning and examine its specificity.

Traditionally, the first geometric phenomena presented have concerned the recognition and study of shapes, but there is also a need to pay attention to the relationship between shapes and their functioning in space. This opens up new fields of research, which also include educational aspects of geometry at the lowest educational levels. In addition, this draws attention to the need to create new theoretical foundations. These, in turn, call for consideration in geometric reasoning and affect both the research related to geometric phenomena other than what has been analysed before and the design of teaching.

Geometry in a patterns environment can lead a child to the philosophy of mathematical thinking:

- from perception to definition and mathematical formulation and
- in finding the general in the particular.

Research into children's reasoning has shown that they are able to perceive a general rule and feel the universality of certain mathematical relationships and that these findings are accompanied by deep emotions. We emphasize here that these tasks lead to conjecture and hypotheses and sensitize the relationships between concepts and teaching how to use these compounds. They provide an opportunity for achieving higher goals in mathematical education: reasoning, communicating mathematically, and using mathematical thinking, both in facts and skills or algorithms. However, when children are able to see a rule or regularity, they must have the opportunity to work in a situation in which these dominate. "Recursive thinking is an extremely strong element of mathematical thinking. However, it is very difficult to develop and feel suggestive of such a rule, if one looks at a single image representing the general rule" (Mason 2003, pp. 9–17). Therefore, rhythms and geometrical regularity act not only as the introduction to geometrical transformations, they are also a field in which to acquire experience associated with capturing regularity and functioning according to a specific rule. As a result, they are a tool to assist the general intellectual development of the child.

One of the new challenges created both by the theoretical framework (Duval 1998) and by the observations carried out in the course of the research is the functional use of language and building the meanings of words and geometric names. Language is one of the tools helping the transition to a higher level of geometric understanding. Previous studies have confirmed its functionality in this regard. This happens regardless of the research domain in geometry and applies to both understanding shapes and geometric relationships. This fact also affects the methodology of the conducted study. At the lowest levels of creation of geometrical concepts, visual language is essential. An attempt to switch to other types of expression repeatedly forces the language of gestures rather than verbal expression. These observations are the basis for a teaching design in which both the visual information and gestures are insufficient means of communication.

Geometry for the youngest is not an enclosed teaching area. It is a first step in gaining general geometrical knowledge (students' mathematical knowledge). Therefore, the didactic and psychological problems of small children's creation of geometrical concepts must be seen with a wide perspective.

References

Arcavi, A. (2003). The role of visual representations in the learning of mathematics. *Educational Studies in Mathematics*, 215–241.
Arzarello, F. (2004). Mathematical landscapes and their inhabitants: Perceptions, languages, theories, Plenary lecture, ICME 10, Copenhagen.
Bartolini Bussi, M. G., & Mariotti, M. A. (2008). Semiotic mediation in the mathematics classroom: Artifacts and signs after a Vygotskian perspective. In L. English et al. (Eds.), *Handbook of international research in mathematical education* (pp. 746–783). Routledge.

Battista, M., & Clements, D. H. (1996). Finding the number of cubes in rectangular cube buildings. *Teaching Children Mathematics, 4*(5), 258–264.

Boero, P., & Consogno, V. (2007). Analyzing the constructive function of natural language in classroom discussions. In D. Pitta-Pantazi, G. Philippou (Eds.), *Proceedings CERME 5* (pp. 1150–1159).

Brousseau, G. (1983). Les obstacles épistémologiques et les problèmes en mathématiques. *Recherches en Didactique des Mathématiques,* 165–198.

Brunheira, L., & da Ponte, J. P. (2015). Prospective teachers' development of geometric reasoning through an exploratory approach. In K. Krainer & D. Vondrov'a (Eds.), *CERME 9—Ninth congress of the European society for research in mathematics education*, Feb 2015, Prague, Czech Republic. (Proceedings of the Ninth Congress of the European Society for Research in Mathematics Education.hal-01287007).

Budden, F. J. (1972). *The fascination of groups.* Cambridge: Cambridge University Press.

Bulf, C. (2010). The effects of the concept of symmetry on learning geometry at French Secondary School. In *Proceedings CERME 6* (pp. 726–735).

Bulf, C., Marchini, C., & Vighi, P. (2013). Le triangle-acrobate: Un jeu géométrique sur les isometries en CE1. *Intérêts et limites,* Grand N, 91, Irem, Grenoble, 43–70.

Bulf, C., Marchini, C., & Vighi, P. (2014a). Analisi di un gioco sulle isometrie nella scuola primaria: il triangolo-acrobata. *L'Insegnamento della Matematica e delle Scienze Integrate,* 37°, 1, Paderno del Grappa (TV): Centro Morin, 7–33.

Bulf, C., Marchini, C., & Vighi, P. (2014b). Preconcetti sulle isometrie nella scuola primaria. Un case-study condotto in Francia e in Italia. *L'Insegnamento* della *Matematica e delle Scienze Integrate,* 37A, 2, Paderno del Grappa (TV): Centro Morin, 107–132.

Cilavdaroğlu, A. K. (2012). *İlköğretim matematik öğretmenliği birinci sınıf öğrencilerinin bazı iki boyutlu geometrik kavramların tanımları ve şekillerine dair bilgilerinin incelenmesi, Yayınlanmamış yüksek lisans tezi,* Gaziantep Üniversitesi, Sosyal Bilimler Enstitüsü.

Clarke, B., Clarke, D., & Cheeseman, J. (2006). The mathematical knowledge and understanding young children bring to school. *Mathematics Education Research Journal, 18*(1), 78–102.

Clements, D. H. (2001). *Mathematics into preschool, teaching children mathematics.* http://gse.buffalo.edu/fas/clements/files/Preschool_Math_in_TCM.pdf

Clements, D. H., & Battista, M. T. (1992). Geometry and spatial reasoning. In D. A. Grouws (Ed.), *Handbook of research on mathematics teaching* (pp. 420–464). N.C.T.M.-Macmillan

Clements, D. H., & Sarama, J. (2016). *Young children's conceptualization and learning of geometric figures.* Paper presented to Topic Study Group 4 (TSG4) at the 13th International Congress on Mathematical Education (ICME-13). Hamburg. July 24–31, 2016.

Clements, D. H., Swaminathan, S., Hannibal, M. A. Z., & Sarama, J. (1999). Young children's concepts of shape. *Journal for Research in Mathematics Education, 30,* 192–212.

Coutat, S., & Vendeira, C. (2016, July). *Shape recognition in early school.* Paper sent for the 13th international congress on mathematical education, Hamburg, Germany.

De Lange, J. (1987). *Mathematics, Insight and Meaning.* Utrecht, The Netherlands: OW&OC.

Denizli, Z. A, Erdoğan, A. & Olkun, S. (2016). *The development of three-dimensionality in primary school children* (ICME-13).

Demidow, W. (1989). *Patrzeć i widzieć.* Wydawnictwo Czasopism i Książek Technicznych NOT-SIGMA, Warszawa.

Devlin, K. (1996). *The science of patterns.* New York: Scientific American Library.

Dorier, J.-L., Gutiérrez, Á. & Strässer, R. (2004). Geometrical thinking. In M. A. Mariotti (Ed.), *Proceedings of the third congress of the European society for research in mathematics education.*

Duval, R. (1993). Registres de représentation sémiotiques et fonctionnement cognitif de la pensée. *Annales de Didactique et de Sciences Cognitives, 5,* 37–65.

Duval, R. (1998). Geometry from a cognitive point of view. In C. Mammana & V. Villani (Eds.), *Perspectives on the teaching of geometry for the 21st century: an ICMI study* (Vol. 5, pp. 37–51). Dordrecht, Boston: Kluwer Academic Publishers.

Duval, R. (2005). Les conditions cognitives de l'apprentissage de la géométrie: développement de la visualisation, différenciation des raisonnements et coordination de leur fonctionnement. *Annales de Didactique et de Sciences Cognitives, 10,* 5–53.

Duval, R. (2006). A cognitive analysis of problems of comprehension in a learning of mathematics. *Educational Studies in Mathematics, 61,* 103–131.

Economopoulos, K. (1998). What comes next? The mathematics of pattern in kindergarten. *Teaching Children Mathematics, 5*(4), 230–234.

Encyklopedia Szkolna, Matematyka (*group work*). (1990). Wydawnictwa Szkolne i Pedagogiczne, Warszawa. 383.

Fischbein, E. (1993). The theory of figural concepts. *Educational Studies in Mathematics, 24*(2), 139–162.

Fujita, T. (2010). Understanding of the hierarchical classification of quadrilaterals. In M. Joubert (Ed.), *Proceedings of the British society for research into learning mathematics* (Vol. 28, No. 2). Available at bsrlm.org.uk. June 2008 From Informal Proceedings 28-2 (BSRLM)

Fujita, T., & Jones, K. (2007). Learners' understanding of the definitions and hierarchical classification of quadrilaterals: towards a theoretical framing. *Research in Mathematics Education, 9*(1 & 2), 3–20.

Ginsburg, H. P. (2004). Little children, big mathematics: Learning and teaching in the Pre-School. In A. Cockburn & E. Nardi (Eds.), *Proceedings of the 26th conference of the international group for the psychology of mathematics education* (Vol. 1, pp. 3–14). Norwich, UK: PME.

Gonseth, F. (1945–1955). *La géométrie et le problème de l'espace.* Lausanne: Le Griffon.

Grabowska, A., & Budohoska, W. (1992). Procesy percepcji. In T. Tomaszewski (Ed.), *Psychologia ogólna.* Warszawa: Wydawnictwo Naukowe PWN.

Gray, E., & Tall, D. (1994). Duality, ambiguity and flexibility: A proceptual view of simple arithmetic. *Journal for Research in Mathematics Education, 25*(2), 116–140.

Gutiérrez, A., & Jaime, A. (1998). On the assessment of the Van Hiele levels of reasoning. *Focus on Learning Problems in Mathematics, 20*(2/3), 27–46.

Gutierrez, A., Jaime, A., & Fortuny, J. M. (1991). An alternative paradigm to evaluate the acquisition of the Van Hiele levels. *Journal for Research in Mathematics Education, 22*(3), 237–251.

Hejný, M. (1993). The understanding of geometrical concepts. In *Proceedings of the 3rd Bratislava international symposium on mathematical education, BISME3.* Bratislava: Comenius University.

Hejný, M. (2012). Exploring the cognitive dimension of teaching mathematics through scheme-oriented approach to education. *Orbis Scholae, 6*(2), 41–55. http://www.orbisscholae.cz/archiv/2012/2012_2_03.pdf

Hejný, M., Jirotková, D., Slezáková, J., Bomerová, E., & Michnová, J. (2007–2011). *Matematika pro 1.-5. ročník ZŠ. Učebnice a Příručka učitele.* Plzeň: Fraus.

Hsu, W.-M., & Lin, M.-J. (2009). A content analysis of geometry materials in elementary mathematics textbook of Taiwan, China and Hong Kong. *Journal of Education National Changhua University of Education, 16,* 47–63.

Jagodzińska, M. (1991). *Obraz w procesach poznania i uczenia się. Specyfika informacyjna, operacyjna i mnemiczna.* Warszawa: WSiP.

Jones, K., & Mooney, C. (2003). Making space for geometry in primary mathematics. In I. Thomson (Ed.), *Enhancing primary mathematics teaching* (pp. 3–15). London: Open University Press.

Jirotková, D. (2010). *Cesty ke zkvalitňování výuky geometrie.* Praha: UK v Praze, Pedagogická fakulta, 330 p. ISBN 978-80-7290-399-3.

Jirotková, D. (2011). Generic models in geometry. In J. Novotná, & H. MORAOVÁ (Eds.), *Proceedings of the SEMT '11—International symposium, elementary maths teaching.* Praha: UK v Praze, PedF, s. 174–181. ISBN 978-80-7290-502-6.

Jirotková, D. (2016). *Scheme of geometrical concepts.* Paper presented to Topic Study Group 4 (TSG4) at the 13th international congress on mathematical education (ICME-13). Hamburg. July 24–31, 2016.

Kaufman, L. (1979). *Perception. The world transformed.* New York: Oxford University Press.
Kloboučková, J., Jirotková, D. & Slezáková, J. (2013). Enhancement of 3-D imagination in the 1st and 2nd Grade. In *Procedia—Social and Behavioral Sciences* (Vol. 93, pp. 984–989). October 21, 2013.
Konior, J. (2003). Studium płaszczyzny—dydaktyczne i psychologiczne aspekty rozwoju pojęcia w kontekście uczniowskiej percepcji rysunku geometrycznego w planimetrii. *Studia Matematyczne Akademii Świętokrzyskiej* 10 (pp. 139–163). Kielce: Wydawnictwo Akademii Świętokrzyskiej.
Kordos, M. (2005). *Wykłady z historii matematyki.* Warszawa: SCRIPT.
Kuzniak, A., & Houdement, C. (2001). Elementary geometry split into different geometrical paradigms. In M. A. Mariotti (Ed.), *Proceedings of the third congress of the European society for research in mathematics education.*
Kuzniak, A., & Nechache, A. (2015). Using the geometric working spaces to plan a coherent teaching of geometry. In K. Krainer & N. Vondrov'a (Eds.), *Proceedings of the ninth congress of the European society for research in mathematics education* (pp. 543–549). *CERME 9.* Feb 2015, Prague, Czech Republic.
Kuřina, F. (1995). The first geometrical experience of a child. In *Proceedings of international symposium of elementary math teaching* (pp. 42–45). Prague: Charles University, August 28– September 1.
Lee, T.-N. (2016). *Using questioning and argumentative activities to help grade 5 students generalize triangle properties.* Paper presented to Topic Study Group 4 (TSG4) at the 13th International Congress on Mathematical Education (ICME-13). Hamburg, Germany. July 24– 31, 2016.
Levenson, E., Tirosh, D., & Tsamir, P. (2011). *Preschool geometry, theory, research, and practical perspectives.* Rotterdam/ Boston/ Taipei: Sense Publishers.
Mack, N. K. (2007). Gaining insights into children's geometric knowledge. *Teaching Children Mathematics, 14*(4), 238–245.
Marchini, C. (2004). Different cultures of the youngest students about space (and infinity). In M. A. Mariotti (Ed.), *Proceedings CERME 3* (pp. 1274–1284). Bellaria.
Marchini, C., & Vighi, P. (2005). The richness of patterns. Applications to mathematics education in Italy. In P. Palhares (Ed.), *Proceedings EME 2004.* Braga: Universitade do Minho.
Marchini, C., & Vighi, P. (2007). Geometrical tiles as a tool for revealing structures. In D. Pitta-Pantazi & G. Philippou (Eds.), *Proceedings CERME 5* (pp. 1032–1041). Larnaca.
Marchini, C., & Vighi, P. (2009a). Can we develop geometrical understanding by focusing on isometries? A teaching experiment by the means of geometrical artefacts. In J. Novontá (Ed.), *Proceedings of international symposium elementary maths teaching* SEMT '09 (pp. 169–176). Prague.
Marchini, C., & Vighi, P. (2009b). Can we develop geometrical understanding by focusing on isometries? A teaching experiment by the means of geometrical artefacts. In J. Novontá (Ed.), *Proceedings of international symposium elementary maths teaching* SEMT '09 (pp. 169–176). Prague: Czech Republic.
Marchini, C., & Vighi, P. (2011). Innovative early teaching of isometries. In M. Pytlak, T. Rowland, & E. Swoboda (Eds.), *Proceedings CERME7* (pp. 547–557). Poland: University of Rzeszów.
Marchini, C., Swoboda, E., & Vighi, P. (2008). How to reveal geometrical independent thinking in the lower primary years. In B. Maj, M. Pytlak, & E. Swoboda (Eds.), *Supporting independent thinking through mathematical education* (pp. 82–88). Wydawnictwo U.R: Rzeszów.
Marchini, C., Swoboda, E., & Vighi, P. (2009). Indagine sulle prime intuizioni geometriche mediante 'mosaici'. Un esperimento nella scuola italiana. *La Matematica e la sua didattica, 23* (1), 61–88.
Mason, J. (2003) Structure of attention in the learning of mathematics. In J. Novontá (Ed.), *Proceedings of international symposium elementary maths teaching* SEMT '03 (pp. 9–17). Prague, Czech Republic.

MPI (Ministero della Pubblica Istruzione). (2007). *Indicazioni per il curricolo per la scuola dell'infanzia e per il primo ciclo d'istruzione.* Roma.

MPI (Ministero della Pubblica Istruzione). (2012). *Indicazioni per il curricolo per la scuola dell'infanzia e per il primo ciclo d'istruzione.* Roma.

Nùñez, R. E., Edwards, L. D., & Matos, J. P. (1999). Embodied cognition as grounding situatedness and context in mathematics education. *Educational Studies in Mathematics, 39*(1–3), 45–65.

Ozdemir, E., & Dur, Z. (2014). Preservice mathematics teachers' personal figural concepts and classifications about quadrilaterals. *Australian Journal of Teacher Education, 39*(6), 107–133. http://dx.doi.org/10.14221/ajte.2014v39n6.1

Piaget, J., & Inhelder, B. (1947). *La réprésentation de l'éspace chez l'enfant.* Paris: PUF.

Reinhold, S., & Wöller, S. (2016). *Children's conceptual knowledge on cubes and cuboids: insights via block-building activities.* Paper presented to Topic Study Group 4 (TSG4) at the 13th International Congress on Mathematical Education (ICME-13). Hamburg, Germany. July 24–31, 2016.

Rożek, B., & Urbanska, E. (1998). Children's understanding of the row-column arrangement of figures. In F. Jaquet (Ed.), *Relationship between classroom practice and research in mathematics education, Proc. CIEAEM 50* (pp. 303–307), Neuchâtel.

Satlow, E., & NewCombe, N. (1998). When is a triangle not a triangle? Young children's developing concepts of geometric shape. *Cognitive development, 13*(4), 547–559.

Sękowski, T. (1996). *Człowiek i matematyka.* Warszawa: Polska Oficyna Wydawnicza "BGW".

Shannon, C. E., & Weaver, W. (1949). *The mathematical theory of communication.* Urbana (US): University of Illinois Press.

Sinclair, N., & Bruce, C. D. (2015). New opportunities in geometry education at the primary school. *ZDM, 47*(3), 319–329.

Speranza, F. (1997). *Scritti di Epistemologia della Matematica.* Bologna: Pitagora.

Steen, L. A. (1990). *The science of patterns.* Science, *29.*

Swoboda, E. (2005). Structures of Van Hiele's visual level in work of 5–7 years old children. In J. Novontá (Ed.), *Proceedings of international symposium elementary maths teaching SEMT '05* (pp. 299–306).

Swoboda, E. (2006). *Przestrzeń, regularności geometryczne i kształty w uczeniu się i nauczaniu dzieci.* Rzeszów (Poland): Wydawnictwo Uniwerytetu Rzeszowskiego.

Swoboda, E. (2007). Intuition of geometrical relations in plane in works of 4–7 year old children. In J. Novontá, & H. Moraová (Eds.), *Proceedings of the international symposium elementary maths teaching SEMT '07* (pp. 249–257).

Swoboda, E. (2011). Axis symmetry as an epistemological obstacle. In J. Novotná, & H. Moraová (Eds.), *The mathematical knowledge needed for teaching in elementary schools. Proceedings SEMT '11* (pp. 320–328). Prague: Charles University, Faculty of Education.

Swoboda, E. (2013). Investigating manipulations in the course of creating symmetrical pattern by 4–6 year old children. In B. Ubuz, C. Haser, & M. A. Mariotti (Eds.), *Proceedings of the eight congress of the European society for research in mathematics education* (pp. 685–694). Ankara: Middle East Technical University. ISBN 978-975-429-315-9.

Swoboda, E., & Tatsis, K. (2010). Five-year-old children construct patterns, deconstruct them and talk about them. *Annales of the Polish Mathematical Society, 5th Series: Didactica Mathematicae, 32,* 153–174

Szemińska, A. (1991). Rozwój pojęć geometrycznych (Development of geometrical concepts). In Z. Semadeni (Ed.), *Nauczanie Początkowe Matematyki, Podręcznik dla nauczyciela.* t.1. Wydanie drugie zmienione. Warszawa: WSiP.

Tall, D. (1995) Cognitive growth in elementary and advanced mathematical thinking. *Proceedings of conference of the international group for the psychology of learning mathematics* (Vol. I, pp. 161–175), Recife, Brazil, July 1995.

Tall, D. (2001). What mathematics is needed by teachers of young children? In J. Novontá & H. Moraová (Eds.), *Proceedings of the international symposium elementary maths teaching SEMT 01.* Czech State: Prague.

Trilling, J. (2001). *The language of ornament*. London: Thames & Hudson.

Tsamir, P., Tirosh, D., Levenson, E., Barkai, R. & Tabach, M. (2015). Early-years teachers' concept images and concept definitions: triangles, circles, and cylinders. *ZDM Mathematics Education, 47*, 497–509.

Türnüklü, E., Akkaş, E., & Gündoğdu Alayli, F. (2013). Mathematics teachers' perceptions of quadrilaterals and understanding the inclusion relations. In B. Ubuz, C. Haser & M. A. Mariotti (Eds.), *Proceedings of the eight congress of the European society for research in mathematics education* (pp. 705–714).

Usiskin, Z., & Griffin, J. (2008). *The classification of quadrilaterals: A study of definition.* Charlotte, NC: Information Age Publishing.

Van Hiele, P. (1986). *Structures and insight. A theory of mathematics education.* London: Academic Press.

Van Heuvel-Panhuizen, M., & Buys, K. (Eds.). (2004). Young children learn measurements and Geometry. TAL Project, Freudenthal Institute, Utrecht University, National Institute for Curriculum Developmnet (SLO) in collaboration with CED Rotterdam.

Verillon, P., & Rabardel, P. (1995). Cognition and artefact: a contribution to the study of thought in relation to instrumented activity. *European Journal of Psychology of Education, 10*(1), 77–101.

Vergnaud, G. (1990). Epistemology and psychology of mathematics education. *Mathematics and cognition: A research synthesis by the international group for psychology of mathematics education ICMI study series* (pp. 14–30). Cambridge: Cambridge University Press.

Vighi, P. (2003a). The triangle as a mathematical object. In M. A. Mariotti (Ed.), *Proceedings of the third congress of the European society for research in mathematics education*. Bellaria (Italy): Università di Pisa.

Vighi, P. (2003b). Pre-conception about triangle. In J. Novotna (Ed.), *Proceedings of SEMT'03* (pp. 152–157), Prague (Czech Republic), ISBN 80-7290-132-X.

Vighi, P. (2006). *'Costruiamo un bel pavimento. Indagine su alcune pre-concezioni e intuizioni relative all'organizzazione spaziale'.* In B. D'Amore, & S. Sbaragli (Eds.), Pitagora (BO), 13–16.

Vighi, P. (2015, in print). Abstract paintings, objects and actions: how to promote geometrical understanding. *Proceedings Mathematical Trangressions II, Krakow.*

Vighi, P., & Marchini, C. (2014). First graders' geometrical speech. The slanting giraffe's neck. In M. Pytlak (Ed.), *Communication in the mathematical classroom* (pp. 108–122). Wydawnictwo Uniwersytetu Rzeszowskiego: Rzeszów.

Viholainen, A. (2006). Relationships between informal and formal reasoning in the subject of derivative. In M. Bosch (Ed.), *Proceedings of the fourth congress of the European society for research in mathematics education* (pp. 1811–1820). Spagna: Sant Feliou de Guixols.

Vopěnka, P. (1989). *Rozpravy s geometrii.* Praha: Panorama.

Walcott, C., Mohr, D., & Kastberg, S. E. (2009). Making sense of shape: An analysis of children's written responses. *The Journal of Mathematical Behavior, 28*(1), 30–40.

Waters, J. (2004). A study of mathematical patterning in early childhood settings. In I. Putt, R. Faragher, & M. McLean (Eds.), Mathematics education for the 3rd millennium: Towards 2010 (565–572). *Proceedings of the 27th Annual Conference of the Mathematics Education Research Group of Australasia*. Sydney: MERGA.

Weyl, H. (1952). *Simmetry.* Princeton University Press.

Yenilmez, K., & Yaşa, E. (2008). İlköğretim öğrencilerinin geometrideki kavram yanılgıları. *Uludağ Üniversitesi Eğitim Fakültesi Dergisi, 21*(2), 461–483.

Yuan, Y. (2016, July). *Effect of different manipulatives on first graders' learning of the ability to count cubic blocks in a 3-D figure.* Paper presented to Topic Study Group 4 (TSG4) at the 13th International Congress on Mathematical Education (ICME-13). Hamburg, Germany, July 24–31, 2016.

www.ingramcontent.com/pod-product-compliance
Ingram Content Group UK Ltd.
Pitfield, Milton Keynes, MK11 3LW, UK
UKHW020216231225
466357UK00011B/170